T0252246

Luminos is the Open Access monograph publishing program from UC Press. Luminos provides a framework for preserving and reinvigorating monograph publishing for the future and increases the reach and visibility of important scholarly work. Titles published in the UC Press Luminos model are published with the same high standards for selection, peer review, production, and marketing as those in our traditional program. www.luminosoa.org

The Scarcity Slot

The Scarcity Slot

Excavating Histories of Food Security in Ghana

Amanda L. Logan

UNIVERSITY OF CALIFORNIA PRESS

University of California Press
Oakland, California

Suggested citation: Logan, A. L. *The Scarcity Slot: Excavating Histories of Food Security in Ghana*. Oakland: University of California Press, 2020.
DOI: https://doi.org/10.1525/luminos.98

Library of Congress Cataloging-in-Publication Data

Names: Logan, Amanda L., author.
Title: The scarcity slot : excavating histories of food security in Ghana / Amanda L. Logan.
Other titles: California studies in food and culture; 75.
Description: Oakland, California : University of California Press, [2020] | Series: California studies in food and culture; 75 | Includes bibliographical references and index.
Identifiers: LCCN 2020022842 (print) | LCCN 2020022843 (ebook) | ISBN 9780520343757 (paperback) | ISBN 9780520975149 (ebook)
Subjects: LCSH: Food security—Ghana—Banda (Brong-Ahafo Region)
Classification: LCC HD9017.G453 B45 2020 (print) | LCC HD9017.G453 (ebook) | DDC 338.1/9667—dc23
LC record available at https://lccn.loc.gov/2020022842
LC ebook record available at https://lccn.loc.gov/2020022843

Manufactured in the United States of America

27 26 25 24 23 22 21 20
10 9 8 7 6 5 4 3 2 1

For the *cholƐƐlƐƐs* of Banda

CONTENTS

ILLUSTRATIONS

MAPS

FIGURES

TABLES

This book is the product of nearly two decades of learning about and investigating African foodways. Most Westerners assume African food and agriculture are impoverished in some way, a view created by popular media and a lack of knowledge about Africans and their rich histories. But the beauty of empirical research is that with each new source of data we peel away another set of assumptions and begin to reveal new narratives and possibilities.

This volume presents a critical engagement with some of those layers, ones that I term, cumulatively, "the scarcity slot." My primary goal is accessibility, so I try to avoid jargon, scientific names, and academic beating around the bush. The critiques I make are direct, because I feel it is our job as scholars and educators to debunk some of the more problematic stereotypes that have emerged, especially those with real-life implications for the people we work with. Such stereotypes often have long afterlives, and changing deeply entrenched assumptions is an uphill battle, as attested by the large array of works devoted to toppling the most tenacious stereotypes about the African continent. I add to these efforts and try to make these engagements clear for students of African food security, food history, and archaeology, as well as policymakers who might be curious about how the past informs the present.

I have been exceptionally fortunate to work with the Banda community in west-central Ghana, and to be able to build upon the foundational research and relationships established there by Ann Stahl. A key frame throughout the volume, and an assumption that is fundamental to anthropology and social history, is the value of deep engagement with local contexts. In North America, archaeologists are usually trained as anthropologists, though that is not always reflected in the questions we ask about the past or in our engagement with local descendent populations. Over the course of many field seasons across East Africa, West Africa, and South America, I became increasingly frustrated with how the questions posed by standard archaeological research often lacked relevance to the rural and often impoverished communities surrounding archaeological sites.

These concerns came to a head when I started to interview elderly women in Banda in 2009. I hoped to understand changes in food practices over time, but I soon learned that food change was intricately bound up in larger political, economic, and cultural shifts. Indeed, food was often the canary in the coal mine, one of the first indicators that major structural changes were afoot. This recognition spurred me to adopt a more open interview format, in the hopes of understanding the context of food change. I made sure my interlocutors could ask me questions too. The first and most common question for me was: "Of what use is this project to us?" Although this query had long been brewing in the back of my head, I did not know how to answer it—but I tried, and it was in these efforts that the seeds of this monograph were born. This book is an attempt to answer that question, even if more work remains to be done.

I am lucky to be able to build off (and access, thanks to Northwestern University's exceptional African Studies Library) the work of scholars across a wide array of disciplines including archaeology, anthropology, history, agrarian studies, and more. The topic of my research, rather than the discipline in which I am trained, has defined my focus. As Scholliers and Claflin (2012) argue, food history is a nondisciplinary affair. Throughout the volume, I build on the excellent work of scholars in other disciplines—especially African history—and suggest that we use these rich insights to advance a dedicated history of food security in the African continent. The diverse array of scholarship that I am able to pull into the service of creating this history attests to its legibility and quality, but as is to be expected when using studies on one topic to address another, gaps emerge, and it is important that we acknowledge those holes. My goal in this volume is not to attempt a comprehensive exposition but rather to bring historical works related to food security in West Africa together in order to provoke more work in this area.

· · ·

The seeds for this study were planted in my early days as an undergraduate student, when I was fortunate to come under the mentorship of Cathy D'Andrea. Cathy taught me how to analyze plant remains and introduced me to Africa in the classroom and through fieldwork in Ethiopia and Sudan. She also trained me in ethnoarchaeology, a method I expanded into the food ethnography I detail in the final chapters of this book. During master's studies with Debby Pearsall, I not only gained technical aptitude in phytolith analysis, but also, and perhaps just as importantly, learned how empirical data can challenge deeply entrenched beliefs about the past. Christine Hastorf's work and mentorship have encouraged me to explore how paleoethnobotany can be used to speak to social theory. I was fortunate to work on Ann Stahl's Banda Research Project in Ghana for my dissertation research. Not only was the archaeology just plain "cool," but many local community members were invested in talking about the past. The now almost thirty-five-year span of the BRP also meant that we knew a tremendous amount about Banda's past. Under Ann's tutelage, the BRP team collected over sixteen hundred flotation samples from nearly all contexts excavated, as well as a large number of soil and artifact samples, providing a ready archive for me to explore changing foodways. A major turning point and inspiration came during a postdoc in interdisciplinary food studies at Indiana University, as part of a Mellon Sawyer seminar on food choice. Thanks to Rick Wilk, Peter Todd, Sara Minard, and Shingo Hamada, I got a crash course in food studies that gave me the language and concepts to link foodways past and present.

Working with Banda community members has had major impacts on the kinds of questions I ask of the past by challenging me to connect past and present. They are gracious hosts, patient interlocutors, and excellent teachers. The Banda Traditional Council, led by Tolɛɛ Okokyeredom Kwadwo Sito, permitted this research and also provided critical commentary on many of its aspects. I am particularly indebted to Enoch Mensah, who served as my research assistant, translator, facilitator of all things, and friend. Many of his insights are reflected throughout these pages. Queen Mother Akosua Kepefu and the cholɛɛlɛɛs (old ladies) of Banda shared their wisdom and persistence with me and challenged me to make history of use to them, so it seems fitting to dedicate the book to them. Sampson Attah and Afriye provided me a home away from home and the best beans I have ever enjoyed. Kwesi Millah and Afua Nimena's vibrant recollections of the past significantly colored my visions of it.

I also owe a tremendous debt to Ann Stahl, for her commitment to understanding Banda's past as well as the access she offered to Banda's rich archive. It is the understatement of the century to acknowledge that her approach has both provoked and made possible the arguments presented in this book. Her high standards have helped me think bigger and bolder and better, and her mentorship and wisdom are reflected on every page of this book.

The Ghana Museums and Monuments Board granted permissions for all research reported here. I also owe thanks to faculty and students of the Department of Archaeology and Heritage Studies at the University of Ghana who have always gone out of their way to make me feel welcome at Legon, particularly Wazi Apoh, William Gblerkplor, Ben Kankpeyeng, and Derek Watson. Zonke Guddah, a graduate of Legon, joined me in fieldwork in 2014. The Department of Botany and Herbarium, also at the University of Ghana, identified many of my comparative specimens and was helpful with advice, especially Patrick Ekpe. Nana Akwesi Prempeh, his nephew Michael Asante, and their family members have been welcoming, helpful, and excellent company. They are also responsible for the Asantehene passing right in front of me—a (literally) golden experience I will always remember. I must also thank my field compatriots in Ghana in 2008 and 2009—Amy Groleau, Andrew Gurstelle, Abass Iddrissu, Harry Monney, Benjamin Nutor, Devin Tepleski, and David Tei-Mensah, and those who in years previous were part of the Banda Research Project and painstakingly collected the data that became part of this book.

Many of these ideas continued to grow during my graduate degree at University of Michigan, where I learned the art of being a contrarian and also benefited from strong peer support networks. Jeannette Bond's support and friendship were critical. My Africanist peers, including Anne Compton, Cameron Gokee, Daphne Gallagher, Andy Gurstelle, Stephen Dueppen, and Robyn D'Avignon, prompted important questions and new ways of seeing the past. Margaret Wilson's unwavering support and creative mind and spirit helped me solve many of the most difficult puzzles. Kate Franklin's arrival on the scene gave (and continues to give) me the breath of fresh air that I needed to move in the direction of social theory and critical colonialisms. Sarah Oas helped sort the flotation samples. Faculty members John Speth, Lisa Young, Henry Wright, Ray Silverman, and Elisha Renne provided much-needed support at various stages of the dissertation project. A graduate class with historian Gabrielle Hecht exploded my ideas about writing monographs and encouraged me to reach for wider audiences. But I owe the most thanks to Carla Sinopoli for getting me through the doctoral program. Her confidence in my work and incisive but supportive advice have been instrumental. She was also the first to see a book in my hastily written dissertation. My identifications were substantially improved thanks to a Wenner-Gren-funded visit to the African Archaeology and Archaeobotany Work Group at the Institute for Archaeological Sciences, Goethe University, Frankfurt, hosted by Katharina Neumann and Alexa Höhn.

This work has benefited tremendously from the advice of my colleagues at Northwestern. Melissa Rosenzweig, Cynthia Robin, Mark Hauser, Matthew Johnson, and Mary Weismantel have given me unwavering support and flexibility as I worked through this book, and have served as sounding boards throughout the process. Noelle Sullivan has helped tremendously in the broader framing of this book towards Africanist medical anthropology and critical development. David Schoenbrun has given me incisive, critical, and encouraging commentaries that helped smooth out the contours of my argument, and Helen Tilley's support and scholarship continue to set an upper limit for what is possible. Graduate students at Northwestern have also prompted me to think beyond the limits of my subject area, especially in classes on archaeologies, communities, and publics and on the anthropology of food. The food students (Atmaezar Hariara, Bridgette Hulse, Emily Kamm, Paula Maia, Nicolette Mantica, Sophie Reilly, Emily Schwalbe) read and commented on a draft of this book, pointing out areas of improvement and offering support for the overall project. My own advisees, Dil Singh Basanti, Kacey Grauer, Dela Kuma, and Sophie Reilly, have shaped the arguments here through many conversations and meetings about unrelated things, and tolerated all of the times when I needed to hide in my office and write. Conversations with Tara Mittelberg about critical agricultural development challenged me to expand my focus. Both the Department of Anthropology and the Program of African Studies have provided lively intellectual communities that fostered the ideas in these pages, as well as funding for related projects. And many thanks to Northwestern's generous family leave policies, without which I could never have had a baby and written a book at the same time.

Many of the best observations about my work arose during visits to other campuses to deliver talks or take part in seminars, offered by a host of supportive colleagues whose advice has been critical over the years. In particular, I'd like to thank Stacey Camp, Gayle Fritz, Liza Gijanto, Walter Hawthorne, Michelle Hegmon, Ian Hodder, Matthew Knisley, Fiona Marshall, Joe Masco, Jamie Monson, Kathy Morrison, Shanti Morell-Hart, Peggy Nelson, Lisa Overholtzer, Andy Roddick, François Richard, Kate Spielmann, Barb Voss, Sarah Walshaw, Ethan Watrall, and Alice Yao. Joeva Rock's stunning work on GMOs in Ghana inspired me to push my connections between past and present into the future. Amy Trubek's incisive advice has also helped considerably in urging this project along. I also thank my Chicagoland crew (especially Elizabeth and Bryan Fagan, Kate Franklin, Rebecca Graff, and Mary Leighton) for good advice and good suppers over the last many years.

Peer review has also strongly shaped the product you see here. The scarcity slot frame was originally proposed in an article, but anonymous reviewers helped me see that I needed a book-length treatment to substantiate the idea. I owe the most thanks to the two reviewers of the present manuscript, including Scott MacEachern, who provided constructive criticism of an earlier version of the manuscript and also provided an excellent model of engaged scholarship in his book on Boko Haram. I am particularly indebted to Kathryn de Luna, who carefully dissected the internal logic and evidentiary basis of my arguments and provided an incredibly detailed roadmap of how to make the book stronger. At University of California Press, I thank Kate Marshall and Enrique Ochoa-Kaup for shepherding me through this process and encouraging Open Access publication. I also thank an anonymous reader from the press who helped in the final tweaking of the tone of the manuscript. Caroline Knapp provided high-quality copyediting that certainly improved the manuscript, and many thanks to Andrew Christenson for his indexing expertise.

Funding was provided by National Science Foundation grants to Ann B. Stahl (BCS 0751350, BCS 9410726, BCS 9911690) and myself (BCS 1041948), as well as a Wenner-Gren Foundation Dissertation Fieldwork grant (N013044) and an Engaged Anthropology grant. At University of Michigan, I received funding from various sources including the LSA Regents Fellowship, the Rackham Humanities Research Fellowship, the Graham Doctoral Fellowship from the Graham

Environmental Sustainability Institute, the Griffin Fund and Richard Ford Research Fund from the Museum of Anthropology, and two Rackham Graduate Student Research Grants. Funds from Northwestern University, including the Program of African Studies, were also instrumental. I also gratefully acknowledge the Northwestern Open Access Fund as well as the Kaplan Humanities Publication Subvention Award for supporting the Open Access publication of this book.

Last but most of all, my gratitude goes to my immediate and extended family, who have provided endless support and fun along the way and have borne my absences during this book's gestation and birth. Thank you for your sacrifice.

Introduction

To the Bemba, millet porridge is not only necessary, but it is the only constituent of his diet which actually ranks as food . . . . I have watched natives eating roasted grain off of four or five maize cobs under my very eyes, only to hear them shouting to their fellows later, "Alas, we are dying of hunger . . . We have not had a bite to eat all day."

—AUDREY RICHARDS, *LAND, LABOUR, AND DIET IN NORTHERN RHODESIA*, 1939

The importance of American foods in Africa is more obvious than in any other continent of the Old World, for in no other continent, except the Americas themselves, is so great a proportion of the population dependent on American foods. Very few of man's cultivated plants originated in Africa . . . and so Africa has had to import its chief food plants from Asia and America . . . . As for the influence of these crops before 1850, we might hypothesize that the increased food production enabled the slave trade to go on as long as it did without pumping the black well of Africa dry.

—ALFRED CROSBY, *THE COLUMBIAN EXCHANGE*, 1972

Have the people of Africa always starved? Most people have never heard this question posed. So ingrained is the idea of "Africa" as a scarce place that the continent has become synonymous with need in popular thought.[1] Preoccupied by images of hungry children and drought-ravaged landscapes, we have come to *expect* that we will find food insecurity in every time and place in the continent. But even the casual visitor to an African country observes something much different: people surviving despite the odds stacked against them. This book aims to challenge expectations of scarcity by investigating empirical realities of resilience in the culinary and agricultural history of Banda, a region of west-central Ghana that invokes an altogether different narrative of African food security.

The quotes above, both from pioneers in the study of food, index two different imaginaries of African food scarcity and reveal the crux of the challenge ahead. Audrey Richards (1995 [1939]) pioneered the study of foodways from an anthropological perspective, and her work remains an exemplar for its methodological sophistication. These methods allowed her to recognize that maize (corn) was not considered food to the Bemba, even if it met their caloric needs. This observation would be unremarkable except that it concerns Africans, who are often conceptually stripped of the ability to make choices about what they eat. People thought to be teetering on the edge of survival are rarely accorded the "luxury" of choice. This kind of mentality is clear in Alfred Crosby's conception of the Columbian Exchange. Maize, the same plant the Bemba eschewed in favor of indigenous grains, is accorded by him the role of savior in insuring sufficient calories for African bodies. Africans are portrayed as incapable of developing appropriate crops themselves. In contrast to Richards's empirically rich narrative, Crosby's synthesis provides no data to support his suppositions about Africa. We might justify Crosby's misdeeds as a thing of the past—he wrote *The Columbian Exchange* in the early 1970s, although a revised edition was published in 2003—and he was certainly no card-carrying Africanist. Yet many of his assumptions continue to haunt literature on the Columbian Exchange to this day (see chapters 2 and 3).

In this book, I argue that people in Banda were very much capable of feeding themselves in the centuries and millennia before Europeans took interest in the continent, and indeed afterwards, when livelihoods were reconfigured in the wake of Atlantic trade and colonialism. That this is not the story that is usually told about African peoples and places relates to the space between empirical realities and expectations of the continent, where implicit assumptions of a scarce Africa tend to go unquestioned. The goal of this book is to interrogate and ultimately repurpose this space into a critical zone of inquiry into African food history.

To introduce this endeavor, we need to consider *why* scarcity is the dominant paradigm through which we understand African foodways. To some the answer may seem obvious, since the highest prevalence of undernourishment in the world—one in five people—is found in the African continent. But to emphasize this statistic at the expense of the reverse—four in five people are not undernourished—acts to limit recognition of the tremendous ability of people to survive despite widespread poverty.

The tendency to view Africans in a negative light is unfortunately commonplace, and there is a large volume of scholarship that addresses and combats the oftentimes racist roots of these assumptions (e.g., Hammond and Jablow 1970; Mudimbé 1994; Pierre 2012). I build on this scholarship by exploring how these negative stereotypes infuse common (mis)understandings of food history and food security in the continent. Food—or lack thereof—provides one of the most powerful representations of scarce "Africa," for a multitude of reasons. We experience food viscerally, from pangs of hunger to feelings of satiety and flavors of novelty and delight. Food has direct, visible effects on the body, allowing estimates of caloric intake that underpin assumptions about economic status. Food is a requirement for life, and its absence reveals the stark inequalities that continue to exist today despite record high levels of food production on a global scale. Food is not an inconsequential or apolitical subject; what we say about food in print is both bound up in and contributes to misconceptions rooted in negative stereotypes about the continent, with very real consequences for food policy and development today (see chapter 6; Niemeijer 1996).

Understanding Africa's food history is imperative because it affords an interrogation of whether the scarcities observed today have always defined the continent. History helps us understand the ultimate causes of modern food security problems and the strategies used by people in the past to avert food disasters. Addressing these issues requires the use of alternative archives (Comaroff and Comaroff 1992, 34–35) that approach scarcity from the bottom up. Much like Richards's archive, the data I rely on are decidedly local, and include ethnography and oral histories. My study benefits from a deep-time perspective through the use of archaeology's rich record of everyday life. I weave these archives together to explore the food history of Banda, a locality in Ghana that has seen tremendous change and suggests even bigger possibilities. Banda's case study engenders a counternarrative of African food scarcity that ultimately questions how food security is managed on the continent. When outsiders expect to see scarcity, they have a hard time unseeing it. Food history allows us to see possibility and potential, with far-reaching impacts for how we might envision African futures.

THE SCARCITY SLOT

The driving force behind much environmental policy in Africa is a set of powerful, widely perceived images of environmental change . . . So self-evident do these phenomena appear that their prevalence is generally regarded as common knowledge among development professionals in Afric[a] . . . They have acquired the status of conventional wisdom . . . Images of starving children, and the attribution of blame to natural environmental causes, have become an integral part of the way Africa is perceived in the North. They are signposts to the lie of the land: the reasoning behind them is taken for granted and rarely questioned.

—MELISSA LEACH AND ROBIN MEARNS, *THE LIE OF THE LAND*, 1996

Africanists have long concerned themselves with dismantling pervasive and tenacious stereotypes that define much outsider engagement with the continent and its peoples. Stereotypes tend to stand in for what is considered common knowledge in situations where people lack enough data to make an informed argument. These assumptions are often second nature and become implicit parts of peoples' ways of thinking or arguing. Even though common knowledge is often unsubstantiated, it remains difficult to debunk because of its pervasiveness, since ubiquity is often taken as an indication of obviousness (Keim 2014; Leach and Fairhead 1996; see also similar claims by Žižek 2008 regarding structural violence). Far from being self-evident, so-called common knowledge always comes from somewhere. In this section, I seek to understand how such generalizations are produced and reproduced. Zeroing in on knowledge production gives us the opportunity to disrupt the cycle of these remarkably tenacious stereotypes.

I refer to this collected conventional wisdom on African food and agriculture as the "scarcity slot." The scarcity slot both makes visible the process by which stereotypes are (re)produced and calls up a wide array of critiques from anthropology, geography, and history about misrepresentations of African food and agriculture. My position is that by naming these stereotypes and confronting them directly we may be in a better position to dislocate them. I build on Trouillot's (1991) notion of a "slot" to describe the tendency to fit new groups or information into predefined spaces or types. Slots are not constructed based on empirical realities. Their architecture is built and their contents filled through a process of Othering. The basic process involves comparison of the Self to the Other, with an assumption that the Other lacks a certain quality possessed by the Self (Said 2003). The resulting slot architecture functions to keep subject and object in the same structural position in order to "ensure a certain functioning of power" (Escobar 2012, 163). So long as power inequities exist, outsider imaginaries serve those who construct them rather than do justice to on-the-ground realities. Scholars are no exception, as they too are products of the privileges and prejudices of their time. As Trouillot (1991) outlines, anthropologists constructed an Other for the purposes of study through comparison to themselves, and archaeologists have been guilty of the same in Othering past peoples (Cobb 2005), especially in Africa (Lane 2005; Stahl 2005).

The content of slots changes over time as societies emphasize different priorities and as values and geopolitical relations evolve (Trouillot 1991, 33). While the savage slot outlined by Trouillot referenced sixteenth- to early twentieth-century notions of proper behavior and biopolitical relations between Europe and its colonies, in the late twentieth and early twenty-first century the Other has often been constructed according to notions of technological progress and resource availability (Escobar 2012, 144; Ong 1987). Along with water, food is the most essential of all resources, and thus food is central to the construction of the scarcity slot.

Throughout this book, I will critically examine three main tenets of the scarcity slot through empirical examination of culinary history in Ghana. These tenets are rooted in stereotypes that rely on assumptions of scarcity and build on earlier representations of "Africa" as a savage place. First, food is most often portrayed in reference to a *lack* of food, rather than in reference to local tastes or to the rich food traditions that have developed across the continent (Carney and Rosomoff 2009; La Fleur 2012). This mirrors Trouillot's (1991) observation that savages were often stripped of all manners and polite custom, which were conceptualized as the unique domain of Western elites. The second generalization is that African farmers tend to be portrayed as incapable of controlling seemingly hostile environments and as lacking agricultural expertise (Carney and Rosomoff 2009; La Fleur 2012; Richards 1985), making them especially vulnerable to environmental change (Ribot 2014); this is not dissimilar to the portrayal of savages as closer to nature (Trouillot 1991). Finally, African foodways and agriculture are often depicted as timeless and rooted in an unchanging past (Stump 2010). Savages too were placed in an earlier rung of the evolutionary ladder, displacing time and denying coevalness (Fabian 2002). In the remainder of this section, I expand on these assumptions and describe how this book aims to challenge each one.

Tenet 1: Lack of Food

> During the Atlantic slave trade, Europeans knew where cereals were grown in West Africa as well as where to find food surpluses. But over the ensuing centuries of slavery, this knowledge was overlooked and seemingly lost; the enslavement of Africans dehumanized its victims and disparaged their achievements in agriculture and technology. The indigenous African cereals were viewed as nothing more than a few "inferior and miserable food staples."
>
> —JUDITH CARNEY, BLACK RICE, 2001

Today, "Africa" is more often known for its alleged lack of food rather than for its bountiful agricultural resources. As Carney argues, this is a relatively recent view that is closely intertwined with pejorative views of Africans as inferior that emerged to justify the slave trade. In Black Rice, she documents how achievements in African rice cultivation were often attributed to Europeans, erasing centuries of knowledge and skill that were fundamental to the successful rice-based economies of West Africa as well as of the American South. Her exceptional work reveals how the Othering process can create vastly different views of a place or people to suit the needs of Europeans. Early European traders saw an abundance of food in Africa's ports; these observations were central to filling their very real needs for food but also to an image of Africa as plentiful and full of opportunities. As the Atlantic slave trade peaked in the eighteenth century, recasting Africans as inferior was a necessary worldview, and one that extended to their food and agricultural practices.

The content of these stereotypes has continued to evolve. Lack has been an important part of how the world's development experts have understood African food insecurity since at least the 1970s. The Sahel drought of 1968–1972 led to widespread famine and focused global attention on West Africa's food security needs. As Watts (2013) details, the crisis literature focused on lack of food supplies as the primary cause of famine. Experts linked food scarcity to increased risk of political instability under the emerging food security paradigm. Many of the resulting initiatives made a simple correlation between available food supplies and the prevalence of hunger. The solution seemed simple: produce more food through technological improvements. In the following decades, Africanist geographers and anthropologists pointed out the limitations of such a simple correlation. For example, Watts (2013) illustrates that chronic, seasonal hunger is much more prevalent than crisis events like famine. And Berry (1984) points to the difficulty in obtaining even basic estimates of agricultural production. If food supply cannot be measured, it is difficult to blame lack of food for Africa's food security woes, and this problem is compounded by the need to recognize smaller-scale but recurrent seasonal shortages (Watts 2013). Despite this empirical limitation, the idea that Africa needs to produce more food remains remarkably tenacious.

In the early 1980s, Amartya Sen used history to challenge the idea that food shortage was to blame for famine in India. His (1981) historical analyses of famines showed that in most cases, adequate food supplies were available, but were not distributed to those in need. Sen argued that the ultimate causes of famine have more to do with entrenched poverty, which is a historical product rather than a constant (chapter 1). The case study that is the focus of this book confirms this pattern. In the fifteen to seventeen centuries, high levels of food security were maintained even during the worst drought in the last millennium, thanks to a strong local economy (chapter 2). It was only under conditions of privation born out of an emerging colonial global economy that chronic food insecurity emerged in the nineteenth century (chapter 4).

People perceived of as food insecure are often conceptually stripped of their ability to choose what is eaten, so it is not surprising that African tastes and food traditions are less often the subject of serious food study. More is written about French cuisine than the cuisines of all of Africa, a continent that is fifty-five times larger than France. The elevation of particular European cuisines, French chief among them, is rooted in colonial politics of value (Janer 2007). Jack Goody's (1982) *Cooking, Cuisine, and Class* inadvertently exacerbated the problem by making an explicit comparison between African food and that of Europe, effectively Othering African foodways into a non-European slot (see chapter 5). His main argument was that Africans lacked the *haute* cuisines found across Europe because they also lacked similarly pronounced class stratification. This thesis built on earlier work that argued that the different surplus potentials of African and Eurasian agriculture lead to more strongly stratified societies in the latter (Goody 1977). While this type of scholarship has fallen out of favor in anthropology, many of its assumptions and basic problems continue to beleaguer other fields, particularly economic history (e.g., Mayshar et al. 2015). As Africanists, the long afterlife of studies like Goody's should give us pause; while some of his work is still relevant, we have a responsibility to actively critique some of its more problematic aspects and offer alternative narratives. James McCann, the leading voice in African food history, attempts just that in his 2009 book, where he directly confronts Goody's stereotype of African foodways as mundane and undifferentiated by highlighting the range of culinary cultures across the continent as well as the skills of African women as cooks.

This book continues McCann's quest to illustrate the diversity and ingenuity of foodways across the continent, but takes aim specifically at assumptions of lack of food supplies. The perceived scarcity of modern-day African foodways is often extended to our understanding of past food choices, particularly in times and places when empirical data is scant. In this book, one of my major foci is the Columbian Exchange, long of interest to archaeologists and historians but less well attested in empirical records. The Columbian Exchange involved the movement of plants, animals, people, and diseases between hemispheres in the centuries following Columbus's arrival in the Americas, and represents one of the most significant changes in foodways in modern history (Crosby 2003). The introduction of the American staples maize and cassava to Africa has received a great deal of attention, since both crops are critical to maintaining food security in the present (e.g., Alpern 1992, 2008; McCann 2005; Miracle 1966). One hypothesis that has had remarkable staying power is that the introduction of maize made up for population losses suffered during the trans-Atlantic slave trade, since maize could produce higher yields and thus feed greater numbers of people. Originally formulated in the 1960s by Crosby ([1972] 2003) and Curtin (1969), this hypothesis is based on highly suspect data and assumptions (see chapter 3). Yet theses like this endure, particularly outside of anthropology and history (e.g., Cherniwchan and Moreno-Cruz 2019), perhaps because they fit neatly into racialized assumptions about African resources and capabilities.

Recent Africanist food history has offered a different view of the Columbian Exchange by highlighting diverse African agencies during these tumultuous centuries. For example, scholars have shown how new foods like maize were actively employed in strategic ways by farmers and traders along the African coast (Carney and Rosomoff 2009; La Fleur 2012). I extend these works by arguing against assumptions of passivity that are central to the idea that Africans were in need of new foods, as the quote from Crosby that opened this chapter suggests. In Chapters 2–4, I attempt to evaluate whether such a need existed by looking at the adoption of American crops like maize. Like Carney (2001), I also recenter the important roles of African domesticates during the Columbian Exchange, by focusing on the contributions of pearl millet and sorghum, which have largely been overlooked by scholars focusing on this period.

Constructing an "Africa" that has always lacked food and food choices creates a terrain ripe for the expansion of development projects that assume lack of food is the problem to be solved (cf. Mitchell 2002). This worldview perpetuates the view that "Africa" needs outside help to solve its problems, and overlooks the home-grown solutions that may provide more sustainable ways forward. Used critically, the past provides a great potential source of data on strategies that helped people to weather environmental change in the past (Lane 2015; Logan et al. 2019). In chapter 2, I illustrate how local crops helped people persevere during the worst drought on record in the last millennium, and in chapter 6, I consider how these success stories can be used to reset the possibilities for the future.

Tenet 2: Limited by Hostile Environments

Knowledge of food-production [sic] and metals permitted concentrations of population, but slowly, for, except in Egypt and other favoured [sic] regions, Africa's ancient rocks, poor soils, fickle rainfall, abundant insects, and unique prevalence of disease composed an environment hostile to agricultural communities.

—JOHN ILIFFE, *AFRICANS: THE HISTORY OF A CONTINENT*, 1995

Their history was not just a shapeless mass of peoples' comings and goings, helpless before the vagaries of the environment.

—DAVID SCHOENBRUN, *A GREEN PLACE, A GOOD PLACE*, 1998

A strong environmental determinist bent has long permeated outsider approaches to Africa, with prominent scholars arguing that depauperate African environments explained the underdevelopment of the continent (e.g., Iliffe 1995; see Mandala 2005, 10 for a review; see chapter 3). Historians like David Schoenbrun responded by crafting well-supported rebuttals that show cases of clear agricultural abundance, innovation, and resilience. Key in these efforts is inserting African agencies into the equation, since environmental determinist formulations presuppose that African farmers were unable to overcome their inherited environmental limitations. Stereotypes like these have deep roots in racialized assumptions about the continent and its peoples, and take many forms, from assumptions about landscape degradation (Fairhead and Leach 1996; McCann 1999; see chapters 1 and 3) to assumptions about the roles of shifting agriculture and climate change (see excellent critiques by Berry 1993; Moore and Vaughan 1994; Ribot 2013; Richards 1985). At the root of these stereotypes is a process of Othering whereby European landscapes, agricultural practices, and environmental logics are compared to African ones.

Shifting cultivation is one of the clearest examples of this kind of Othering. Also known as slash-and-burn, the practice has been viewed with derision since at least the colonial era (and likely before, see chapter 3). This type of agricultural system involves clearance and burning of vegetation, farming the resulting plot for a year or two, and then moving to new plot until the former one regains fertility as natural vegetation regrows (Richards 1985, 49). To Europeans, shifting cultivation was wasteful since it required large quantities of land per unit population unlike the permanent, intensive fields they relied on at home (Moore and Vaughan 1994). These attitudes can be traced through some of the most influential scholarship on agricultural economics and more broadly, where they were assigned evolutionary significance. For example, Boserup (1965) saw the development of intensive systems as a function of population pressure, which she described as the driving force behind agricultural innovation. To her, shifting cultivation represented a surviving practice of a previous, less populous era (Richards 1985, 52–53).

At the root of this argument is the idea that population pressure and intensive agriculture represent universal progress (Richards 1985, 53). While intensive agriculture uses less land, it is far more demanding of labor. In the regions of tropical West Africa first encountered by Europeans, land was abundant and labor in short supply (Hopkins 1973), so there was no need to intensify production. We cannot assume that these dynamics have characterized the entire continent for all of time; rather, the land/population relationship needs to be approached as an empirical question. The trans-Saharan slave trade likely had a significant impact on portions of West and Central Africa, some of which were depopulated and therefore decreased the availability of labor and skill for agriculture (Carney and Rosomoff 2009; Inikori 1982). Indeed, European chroniclers noted labor shortages along the Gold Coast that were severe enough to create demand for imported slaves from elsewhere in the continent (La Fleur 2012; Lovejoy 2011). In these instances, there was no need for intensive agriculture in tropical West Africa; it was certainly not a "better" option, a perspective shared by the modern food movement that emphasizes low-tech, organic agriculture that is more environmentally sustainable than intensive, industrialized approaches.

Despite these cogent critiques, intensive agriculture (particularly of maize) continues to be regarded as the pinnacle of agricultural achievement among some food historians and archaeologists. The reasoning is that intensive agriculture enables the production of surplus, which can be used to support larger populations and accumulate wealth, which in turn enables the rise of civilizations (Goody 1977). Similar reasoning has been applied to explain the adoption of crops as part of the Columbian Exchange and the consequences of that adoption. Many Africanist historians and archaeologists assume that high-yielding crops like maize were adopted rapidly and enabled population growth, with some going so far as to credit American crops with the growth of state-level societies like Asante (chapter 3). Although this social evolutionary focus has been critiqued by Africanists for some time (e.g., McIntosh 1999; Stahl 1999a), the idea that intensive agriculture and surplus production were desirable has received less critical attention (but see Marshall and Hildebrand 2002; Richard 2017; Widgren 2017). Yet we should be cautious about applying what is essentially a Western economic rationale—and purported evolutionary benchmark—to precolonial African contexts, especially given the discussion of intensive agriculture above.

Despite these mitigating variables, African agriculture continues to be evaluated against Western yield-centric standards by some economic historians (see chapter 1). A brief example illustrates the problematic assumptions of such approaches. In one of the most recent case studies, Rönnbäck and Theodoris (2019) attempt to empirically evaluate historic agricultural productivity in Senegambia as compared to elsewhere in the world. They conclude that nineteenth-century Senegambian agricultural productivity was much lower than yields in nineteenth-century plantation economies in the United States. Yet the higher agricultural productivity of plantation production was built on the backs of enslaved Africans, many of whom were extracted from Senegambia. Labor has been a limiting factor in African agricultural communities, and this situation was likely worsened, or even created, by the siphoning off of agricultural laborers in their prime to be sold as captives in the Americas. The authors use these productivity gaps to argue that it was more profitable to sell Africans to European slave traders because their labor was more valuable in environments that afforded more productive agriculture. While this argument could certainly be deconstructed from a number of angles, what I want to take away from this discussion is that these kinds of questions were asked at all.

The persistence of arguments like these illustrates how badly we need to provincialize the idea that surplus and accumulation are universal priorities. In *Provincializing Europe*, Chakrabarty (2000) argues that Europe has tended to serve as the standard for the rest of the world, and he tasks scholars to treat it like any other place rather than like the pinnacle of human achievement. By this reasoning, surplus accumulation should be treated as an economic strategy used under a historically specific set of conditions rather than as an evolutionary benchmark (see also de Luna 2016). Many African agricultural strategies focus on risk reduction rather than high yields (Richards 1985). High-yielding intensive agriculture generally focuses on monocrops of singular species. These systems are particularly vulnerable to shortfalls in the specific needs of the focus crop (e.g., rainfall), and to diseases that prey on that species. In contrast, the multicropping strategies employed by many African subsistence farmers effectively disperse risk. "Because West African farmers tended to ride with, rather than override, natural diversity, it was assumed that their techniques were especially 'ancient' and 'primitive'" (Richards 1985, 42). What many European observers failed to grasp was that techniques like intercropping are much more appropriate to tropical environments, as the studies cited above demonstrate. The timing of rainfall onset in particular is highly variable in such environments, and can have devastating impacts on crop harvests. In this context, relying on monocrops would be catastrophic. While authors like Rönnbäck and Theodoris (2019) see these environmental constraints as limiting, an opposite interpretation could be offered. Combined with the ingenuity of African farmers, these constraints have created what is likely a more environmentally sustainable system.

Looking through an alternative lens, many Africanist scholars appreciate that the goals of African farmers may have been much different than those of their European contemporaries, that they have been at times rooted in different conceptions of wealth (Richard 2017; Stephens 2018a, 2018b). In some precolonial

African economies, people and their skills sets were valued above the accumulation of material goods (Guyer and Belinga 1995; but see Stephens 2016). While we must be careful not to apply this model, devised for Central Africa, to the continent as a whole, it does provide an alternate conception of wealth that challenges some of the core assumptions of the dominant Western individual accumulation mentality, suggesting that there are multiple other ways in which people past and present have conceptualized wealth. Importantly for this study, the wealth-in-people model evokes a very different strategy for food production and distribution, suggesting food sharing rather than surplus accumulation. In order to retain skilled followers, leaders would have had to feed them well. The ubiquity of highly developed craft specialists across precolonial African landscapes suggests that at least some locations may have produced enough surplus to feed the nonfarming populace (chapter 2). Regions that once adhered to a "subsistence ethic" that emphasized food distribution and sharing may have faced major changes under the transition to individualizing capitalist economies, with serious impacts on food security (chapter 4; Mandala 2005; Watts 2013).

Despite all of these complicating variables, many modern development projects encourage poor farmers to cultivate monocrops of high-yielding crops (e.g., see description of the One Acre Fund in Thurow 2012). The reasoning is that the increased surplus will allow farmers access to increased supplies of cash, which can then be used to invest in more land, animals, or other assets. This strategy derives from Western preconceptions about African agriculture, as well as from Eurocentric ideals of agriculture and surplus production. It goes against millennia of African farming innovations, including the development of systems that effectively manage environmental perturbations. Ultimately, a focus on high-yielding crops may leave African farmers more vulnerable (Logan 2017). Along with a growing number of Africanist archaeologists (e.g., Logan et al. 2019) I argue that we must look to the strategies farmers used in the past to cope with turbulence in order to develop alternative strategies for the future. Across the continent, Africans have dealt with a myriad of environmental, political, and economic changes, responding in innovative ways to the challenge of making ends meet. Archaeology is well suited to tracing these agricultural innovations and building usable pasts (Lane 2015; Stump 2010). Farmers have long found ways to produce more food when needed through processes of extensification or intensification, examples of which found in the past. Many of these African agricultural innovations were social (Guyer 1984; Logan and Cruz 2014; Stone, Netting, and Stone 1990) and they are usually overlooked by a value scheme that uses materialized technological advancements as benchmarks.

Tenet 3: Timelessness

Much writing about precolonial agriculture in Africa still suffers from "the ethnographic present tense." While reading works by historians, and also archaeologists, it is often difficult . . . to discern between on the one hand assumptions based on ethnographic material, and on the other conclusions based on oral or written history or on archaeological results.

—MATS WIDGREN, "*AGRICULTURAL INTENSIFICATION IN SUB-SAHARAN AFRICA, 1500–1800*," 2017

The idea of "Africa" as static and unchanging is a remarkably tenacious stereotype that archaeologists, anthropologists, and historians have actively combated in the last several decades (e.g., Piot 1999; Stahl 2001; Stump 2010). Assumptions of timelessness continue to be extended to agriculture and foodways in particular (Freidberg 2003, 448; Goody 1982, 33), as Widgren laments. For example, throwaway statements about African farmers "toiling in a time warp, living and working essentially as they did in the 1930s," made by journalist Roger Thurow (2012, xix) appeal to a core set of Western beliefs about Africa and evidence little to no knowledge of past decades, much less centuries. This denial of modernity helps justify external interventions meant to "modernize" farmers stuck in the past. What these assumptions overlook is that the poverty they seek to alleviate through making Africans modern may well be the result of modernity itself (chapters 4–6).

Although we have a plethora of scholarship that attests to Africa's rich agricultural history, African food history is still in its early stages, and we lack a dedicated history of food security (chapter 1). Particularly for the nonspecialist, this gap creates the mirage that nothing has changed over time in terms of food scarcity and food practices (Freidberg 2003, 47–48). In this situation, some scholars "play fast and loose with time" (Stahl 2001, 20–27) referring to a "baseline" Africa to fill in informational gaps (Cobb 2005; Stahl 1993, 2001; Stump 2010; Trouillot 1995; Wylie 1985, 1989). This baseline "is composed of unexamined assumptions and empirical generalizations about a relatively unchanging Other before that threshold" (Cobb 2005, 563). One common baseline falls within the late precolonial or early colonial period (nineteenth to twentieth centuries), when written documentation first became available for many parts of Africa. Yet in the nineteenth century, West Africa was fraught with political troubles linked to shifts in global trade that often turned violent and resulted in food shortages (e.g., La Fleur 2012, 137–44; chapters 3 and 4). As a result, this period is hardly representative of the African past. Instead, knowledge should be based in evidence that is coterminous with the period in question (Stahl 1993, 249; Stahl 2001, 20–39), an approach I adopt when possible throughout this book (chapter 1).

The other baseline, as Widgren's remark suggests, is a somewhat ill-defined "ethnographic present," which might include everything from ethnohistoric and early anthropological sources from the late nineteenth and early twentieth centuries to ethnographic and ethnoarchaeological studies set in recent years. As I will show in subsequent chapters, the twentieth century was a time of dramatic change in food and agriculture, and we can document those changes since written and oral evidence is much more available than it is for prior periods. Widgren expresses frustration with this imagined ethnographic present in the context of his attempt to understand changes in land cover and agricultural strategies over the continent as a whole over the last millennium, as part of large-scale efforts to understand the Anthropocene across the globe.[2] Information on farming in the "ethnographic present" is often uncritically interwoven into how archaeologists and historians characterize past agricultural practices, making it exceedingly difficult to reconstruct empirical agricultural histories. It also creates a false sense of continuity between present and past that begs a question: why is it permissible for scholars to conflate these very different contexts for food and agriculture?

Part of the answer lies in the ongoing production and reproduction of the scarcity slot. The timeless trope remains difficult to dislodge because temporal displacement is so central to how the West thinks of itself vis-à-vis Others. Inhabitants of the scarcity slot are depicted as "premodern," lacking the technology, infrastructure, and financial capacity of the "modern." Fabian (2002, 23) refers to this kind of knowledge production as relying on "Typological time"—a time which is measured not in linear years, but in socioculturally constructed binaries (i.e., traditional/modern). Different geographical locations thus are assigned different temporalities, with those practicing allegedly "traditional" behaviors assumed to be survivals of an earlier time, while only those with "modern" habits are allowed to occupy the present. This tendency is what Fabian (2002) refers to as the denial of coevalness, or contemporaneity, with the so-called moderns. This view assumes that history is unidirectional and progressive, silencing processes of decay and slow violence. The lack of written histories in many of the areas designated as traditional exacerbates this temporal displacement, by creating the mirage that if histories are not written down and easily accessible, they simply do not exist (Stahl 2001; Wolf 1982). These forms of temporal displacement lead to the misconception that scarcity is a permanent state of being rather than a result of historical processes (Logan 2016a; Stahl 1993, 249–50).

One of the major goals of this book is to highlight the flawed evidentiary basis upon which these three major received understandings are based, as well as their origins (see chapter 4). This critique impels a revised narrative of scarcity and African food history. We need a different set of assumptions when approaching culinary history in the continent, some of which have already been outlined by recent work on African food history (Carney and Rosomoff 2009; La Fleur 2012; McCann 2005, 2009). These works forefront African agency and ingenuity, and in so doing, decenter the primary role that Europeans have written for themselves in the history of the continent.

ORGANIZATION OF THE BOOK

In order to develop a counternarrative to the scarcity slot, this book tells a tale of one place and how it has been defined by both scarcity and abundance at different points in its history. Banda, in west-central Ghana (map 1), provides the

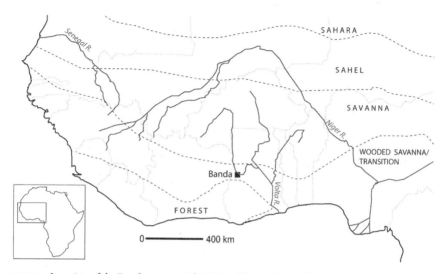

MAP 1. Location of the Banda region within West Africa, showing bioclimatic zones.

ideal case study because it has been the focus of nearly four decades of sustained archaeological and historical research, which has documented the challenges and opportunities afforded by shifting political economic and environmental conditions (Stahl 1999b, 2001, 2007). In 1500, Banda was home to large, bustling towns whose trade tentacles stretched across the premodern world system, and to a strong economy that enabled people to survive the worst drought in a millennium (map 2). Five centuries later, the same town is still vibrant, but occupies a much different position in the world order, and has lost some of the resilience to environmental change observed in the past. Viewing this place over time, we can see potential that multiple forces of history have eroded. This narrative challenges the lens of timelessness and backwardness through which Africa is usually imagined.

In chapter 1, I begin this journey by considering another kind of scarcity that has defined African food history: a lack of traditional source material, which has had a marked impact on how we perceive and write about food scarcity in the past. Food historians studying other locations rely on a creative array of written records as diverse as recipe cards, advertisements, poetry, and ancient imperial policies on adulteration. Unfortunately, this reliance on written sources privileges food stories from parts of the world with long traditions of literacy (e.g., China, Egypt, the Near East, Europe; e.g., Albala 2014). As a consequence, vast areas of the world, including sub-Saharan Africa, are not subject to the kind of critical food history that is needed to debunk stereotypes and build alternative food narratives. Africanist historians and anthropologists have long struggled with the same kinds of challenges, and have developed creative solutions for writing African history that inform my approach to food. Building on these insights, in chapter 1 I advance a *longue durée* history of food security as a corrective to the scarcity slot.

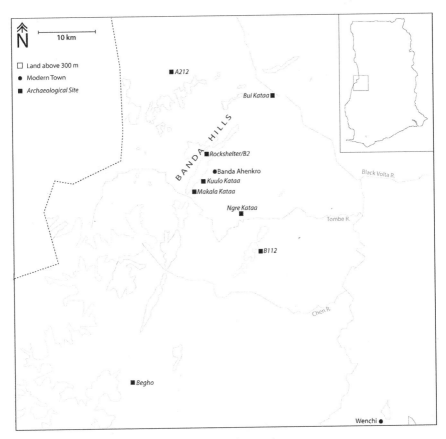

MAP 2. Archaeological sites in the Banda region discussed in text.

While studies of modern-day food security abound, we lack a devoted study of the history of food scarcity and abundance across the continent. From modern studies, we know that food security is not simply a function of how much food is produced, since food supplies are rarely distributed equitably across society. These studies encourage us to pay attention to food access and how it has changed over time. I offer a brief survey of how agricultural, economic, and environmental history hints at major changes in food production and access over the last several thousand years in order to provoke a series of empirical questions for an emergent history of food security over the long term.

Chapter 1 also details my methodological intervention into crafting African food history. I refer to this approach as excavation, building from the basic principles of archaeological excavation. Stratigraphy, or the layers of soil deposited over time, helps us establish chronological control at most sites around the world. We collect artifacts from each layer, comparing them to one another in an attempt to understand contemporaneous events, and to layers above and below in order to grasp change over time. In much the same way, I adhere to a temporal scaffolding to discuss the case study that is central to this book, in order to highlight change over time rather than to apply insights from one period to another uncritically.

The next four chapters tack back and forth between stereotypes about African food writ large and Banda's specific food history over the last six hundred years. This format allows me to illustrate how empirical data from one small place can be used to debunk broader stereotypes about the continent. The stereotypes I choose to critique represent the most pervasive ideas about African food in their respective periods. Rather than trace the genealogies of these ideas—an important task I must leave to future works—I focus on presenting empirical evidence that challenges their core assumptions. This is a deliberate stylistic choice meant to devote the bulk of the book to constructing an alternative narrative based on archaeological and historical evidence and minimizing the oftentimes quite conjectural arguments that have dominated discussions of these periods. In other words, I seek to decenter outsider assumptions and genealogies and center the choices and experiences of African farmers and cooks.

Chapters 2 and 3 are direct critiques of specific misconceptions about the introduction of American crops as part of the Columbian Exchange. In chapter 2, I tackle the notion that Africans were in need of new high-yielding crops from the Americas at the time of their arrival on the continent. To do so, I rely on archaeological data to evaluate the impact of the Columbian Exchange and severe drought in Banda (c. 1400–1650), with emphasis on the introduction of American staple crops like maize. This chapter highlights Banda's potential to weather environmental crises, highlighting the often ignored role of African political economies and indigenous food crops in mitigating drought. The resulting narrative challenges the notion that scarcity has always defined foodways throughout the continent, instead emphasizing culinary preference and choice.

Maize has been accorded a central role in the rise of the Asante state, as well as curative powers in alleviating the low population densities created by the Atlantic slave trade. Using archaeological and historical data, chapter 3 investigates foodways during the height of the Asante empire and Atlantic slave trade, taking the reader from the rich feasts of Asante nobles to the everyday foods of its subjects to the extreme deprivations suffered by those in captivity. This spectrum of food choice highlights the diverse experiences of different African actors during the eighteenth and nineteenth centuries, providing an important counternarrative to views of African food as a simple calories-in-calories-out equation.

Chapter 4 addresses major debates about the impact of colonial interventions on African food security and poverty. Some see colonial interventions as having had little effect on already food-insecure Africans, while others argue that colonial policies utterly transformed Africa's food landscape. I detail Banda's experiences in the late nineteenth and early twentieth centuries. Although not subject to direct agricultural policy interventions, archaeological and oral historical data suggest severe food security challenges during this time, particularly an increasingly severe hungry season. This tumultuous period is often used as a baseline by which to understand African food history, so it is no wonder that scarcity remains an enduring presumption. A *longue durée* view helps us see that this period and its pronounced food insecurity was in many ways an outlier.

In Chapter 5, I examine the relationship between hunger and modernity in postcolonial times (1940s–2009) based on ethnographic interviews. Readers unfamiliar with West African foodways may choose to start with this chapter before reading the historical narrative, to gain an appreciation for the taste preferences and common preparations that define modern African food (or refer to Osseo-Asare 2005, which provides a more comprehensive overview of modern West African food). Instead of providing a primer on African food for the uninitiated, I decided to situate modern foodways in historical space, since they are products of the historical narrative told in the first few chapters of this book. Rather than adopting a chronological focus, chapter 5 highlights the tensions between hunger and modernity in how Banda cooks and farmers make, grow, and abandon foods. Many of these tensions center upon the demands placed on women as cooks, farmers, and mothers, as well as upon the interplay of technology, markets, and aspirations. Rather than using the "ethnographic present" as a means to understand archaeological contexts, I instead consider recent times on their own terms, as products of an ever-changing relationship with the past.

I further explore recent tensions in chapter 6, which opens with an account of the Olden Times Food Fair I organized in Banda in 2014. Heritage foods tend to be viewed negatively in contemporary Banda, because of who makes them and their association with tradition and poverty. These views are rooted not only in local experiences but also in global portrayals of Africa as a scarce, backward place. The forgetting of histories like the one told in this book is rooted in colonial socialities that continue to define modern geopolitical relationships, with real implications for how food security is approached and managed. Remembering those pasts is critical to decolonizing the future.

1

Excavating *Longue Durée* Histories of Food Security in Africa

One of this book's central arguments is that the scarcity slot is constructed and recursively maintained by an apparent scarcity of source material about past foodways. Fairhead and Leach's (1996) field-defining book *Misreading the African Landscape* provides one of the best examples of how the perceived scarcity of sources about West African pasts is an opportunity to bolster stereotypes about African environmental practices. They document how colonial and later administrators and policy-makers interpreted the presence of forest islands as a relic of a once more extensively forested landscape that had been degraded by African farming practices. Through comparing multiple quantitative and qualitative sources, Fairhead and Leach (1996) demonstrate that forest islands were instead created by Africans through generations of careful land management of savannas that would otherwise have lacked such biodiversity. They argue that administrators have essentially read land use history backwards, as indicating deforestation rather than the creation of forest.

Inspired by Fairhead and Leach's approach, I argue that we need to assemble a constellation of sources and methods in order to build *longue durée* histories of African food security. Historians and archaeologists have focused on related and causal issues like agriculture, poverty, development, health, and nutrition, but for the most part, this literature has not been directly marshaled to address the history of food security in the continent (with some important exceptions, e.g., Mandala 2005; Watts 2013). In this chapter, my goal is to bring some of these literatures together in order to advance a set of questions and approaches for conducting a *longue durée* history of food security. I first define food security and discuss how its changing definitions and approaches are relevant to Africa in particular. I briefly outline some of the intellectual influences on the study of the history of food security in Africa. I then explore a number of arguments drawn from archaeology, history, economic history, and anthropology that hint at major changes in African food security while at the same time exposing more questions than answers. In the final section, I detail the methodological approach I take in this book.

DEFINING FOOD SECURITY AND SCARCITY

Concern over having enough of the right food to eat at the right times is older than the human lineage—all organisms structure their lives and activities around the quest to acquire the nutrients and calories necessary for survival. Since Thomas Malthus's highly influential work *An Essay on the Principle of Population* (1798), experts have directed attention to the relationship between the human population and food supplies. Malthus reasoned that population increased geometrically, but that food supplies increased only arithmetically, and concluded that it was inevitable that population would eventually outstrip food supply. When this happened, food shortages, including famine, would reset the balance. He focused particularly on limiting population growth among the poor as a means to prevent wide-scale food scarcity. Though many of his main tenets have received considerable criticism in the last two centuries (e.g., see Devereux 2001; Maxwell 2001), his work continues to have powerful implications, and its main tenets are often subsumed implicitly into how we think about the relationship about food and population growth.

Boserup (1965) leveled an important critique at Malthus's reasoning, showing that in cases of population pressure, people often find ways to increase agricultural production. This has certainly been true of the last half of the twentieth century, in which the tenets of industrialization have been applied to agriculture on a massive scale. Yet the fear that populations will outstrip food supply remains one of the central rallying cries of private and public attempts to increase future food supply. Beginning in the 1960s, neo-Malthusians added the possibility of environmental degradation to Malthus's formulation. Namely, they note that agriculture may produce diminishing returns in situations where soil, for example, has been exhausted. Concern with climate change in recent decades has meant a return to neo-Malthusian ideas, which link environmental degradation to decreased food supply. Since the 1950s, a focus on "resources" has come to replace one on "scarcity," but concern about short supply remains central.

In all of these cases, a lack of food supply, whether caused by population growth or environmental deterioration or climate change, is considered to be the primary source of food insecurity. Yet in the late 1970s and early 1980s, researchers pointed out that food distribution systems are often more critical than food supply to insuring food security. Nobel laureate Amartya Sen's landmark (1981) *Poverty and Famines* argued that food crises are associated with a lack of entitlements (cash, labor, land, etc.) that insure access to food. His argument analyzed historical famines in India and concluded that food supplies during these times were more than sufficient. But the most vulnerable did not have the means to access those foodstuffs, since prices skyrocketed during periods of high demand as colonial officials continued to export grain.

In recent decades, famine analysts have critiqued Sen's overreliance on economic explanations and highlighted the structural, historical processes that create vulnerabilities. They acknowledge political agency as one of the primary causes of famine, shifting blame from nonhuman actors (drought, economies) to agents who might benefit from food crises (see Baro and Deubel 2006; Keen 1994). Other scholars have pointed out that an emphasis on severe and rare events like famines makes it seem as if food insecurity is unusual rather than the product of already existing vulnerabilities (Hendrie 1997; Watts 2013). Rather, they argue, famine is a worst-case manifestation of chronic food insecurity and poverty that leave people vulnerable to environmental, political, or economic shocks.

As a consequence of these and other critiques, there has been an important shift towards measuring and defining access to food rather than simply its availability. Analysts now focus on the household, rather than the nation, as the unit of analysis, which allows a more nuanced view of inequality across space. They also take into account a number of quantitative and qualitative variables—like height/ weight, livelihood, and degree of reliance on coping mechanisms—that permit a more specific examination of household food security and vulnerability (Devereux and Maxwell 2001; see also Wutich and Brewis 2014). Some of these measures are possible for past populations as well, including height and other biometric data from historical and bioarchaeological analyses, household and regional economic data, as well as plant and animal data that speak to coping mechanisms (see Nelson et al. 2016).

However, quantitative metrics do not always adequate capture the perception of hunger, which strongly informs people's actions and feelings of well-being. Quantitative analyses tend to privilege Western, scientific concepts of nutrition and food security over those of local populations who may evaluate hunger and scarcity in very different ways. Consider the two quotes that opened the introduction to this volume, from anthropologist Audrey Richards and environmental historian Alfred Crosby. To Richards, scarcity was in the eye of the beholder. The scene she describes makes clear that the Bemba considered maize as less than food, incapable of filling them up, whereas millet porridge provided true sustenance. Scarcity to the Bemba was experienced as a lack of preferred food, rather than a lack of calories altogether. Juxtapose Richards's take on scarcity against that of Crosby, who might define scarcity on the basis of food production. Crosby wrote at the height of the late 1960s-early 1970s Sahelian famine, when severe drought and unstable governments led to one of the most deadly food crises of the twentieth century. During this era, modernist development experts focused on increasing global food production as a means to insure such famines did not recur. Scarcity, in this view, was something that could be calculated quantitatively, based largely on agricultural production. This national-scale calories-in-calories-out equation is very different than the equation recognized by Richards among the Bemba, of scarcity based on preference. While Crosby's goals were very different than those of the food security analysts discussed here, my point is that we all operate with explicit or implicit ideas of what constitutes and causes food security or scarcity, and these strongly influence the narratives we tell about African foodways.

As Richards's Bemba example demonstrates, the ability to access preferred foods plays a major role in perceptions of scarcity. Scarcity is a feeling of deprivation, whether experienced as physiological hunger pangs or a psychosocially defined deficit. Scarcity is not an absolute lack of resources, but is always defined relative to human activity or social provisioning (Daoud 2010, 1207). Merriam-Webster incorporates both aspects, defining scarcity as "a very short supply; the quality or state of being scarce; especially want of provisions for the support of life." Notice here the incorporation of both a measurable definition, based on quantity or supply, and a more qualitative understanding referring to the state of being scarce or in want. The definition of scarcity used by Merriam-Webster is close to the one I use in this book, because it highlights both quantitative and qualitative aspects, and because these two definitions are often linked. A lack of food resources can easily manifest in feelings of deprivation; the former is often the proximate cause of the latter. But we have to look beyond proximate causation to the structural problems that cause some people to lack resources, if we wish to truly understand scarcity in both guises (Ribot 2014).

Food security is "when all people, at all times, have physical, social, and economic access to sufficient, safe, and nutritious food that meets their food preferences and dietary needs for an active and healthy life" (World Food Summit 1996; see Maxwell 2001 for an overview of the history of the concept). This holistic definition is the one adopted in this book, because it includes attention to multiple aspects or pillars of food security: availability, access, and preference. Availability relates in part to the modernist idea of scarcity, tracking the absolute and quantifiable number of calories produced and available to the populace. Access relates most closely to food distribution, which was the center of Amartya Sen's important intervention and its later refinements. And preference maps onto the kind of scarcity that Audrey Richards calls attention to, the perception of lack because you are unable to eat culturally acceptable or preferred foods. In previous publications (Logan 2016a, 2016b) and throughout this book, I have worked to trace these three constructs. Analyzing how they have changed over time in one small region serves as a backbone to my understanding of scarcity and food security in the archaeological record. My approach differs from that of other archaeologists who study food security (e.g., Nelson et al. 2016) in that I rely on both quantitative and qualitative data, as well as a deep engagement with local historical and ethnographic contexts (see below).[1] In the next section, I explore some of the major themes within and limits to our understanding of African food security and its history writ large. By drawing on qualitative and quantitative studies on related topics like poverty and agriculture, I flesh out a more culturally specific method of exploring food history over time.

THE INTELLECTUAL INGREDIENTS OF
A HISTORY OF FOOD SECURITY

Currently we lack a specific body of scholarship on the history of food security in Africa. But Africanist archaeologists, historians, and geographers have been adept at tracing the closely related topics of agriculture, environment, poverty, and to a lesser extent, food. My work builds on insights from these studies and attempts to combine them to advance an approach to the long-term history of African food security.

Agriculture in particular has received the lion's share of attention because of its central role in colonial and postcolonial African economies. In particular, the so-called agrarian question, whether and how African economies could transition from predominantly agricultural to industrial economies, subsumed much attention in political economy and ecology (see Moyo, Jha, and Yeros 2013), and necessarily focused scholars on production rather than food access or preference. For some scholars, particularly for political ecologist Michael Watts (1983 [2013]), studying famine and food production led to the realization that food distribution and access are among the most critical determinants of hunger. Watts (2013) argued that the incorporation of peasants into global market economies made them much more vulnerable to climatic shifts. His work is a strong influence on interpretations I raise later in this volume, since he deftly uses history to inform our present understandings of food security.

Several authors have responded to stereotypes about African agriculture as backward, with Richards (1985) for example instead demonstrating that farmers actively manipulated their farms to manage risk. A similar preoccupation with overturning assumptions about African environmental management dominated much of the environmental history literature in the 1990s, with Fairhead and Leach's contribution one of the most impactful in terms of policy (see also McCann 1999). Cumulatively, these works push back against the first and second tenets of the scarcity slot—that Africans could not produce enough food and were incapable of modifying their hostile environments—yet for some reason these tenets continue to be maintained, especially in food security and food history literatures. The tenacity of these views can be partly explained by the lack of a dedicated history of food security that explores the complex interplay of food availability, access, and preference over time. This book attempts to provide that, with a focus on the long-term that helps not only to combat the idea that African foodways are timeless, but also to reimagine different futures based on past accomplishments. Archaeologists have demonstrated many of these capabilities by documenting the independent invention of agriculture in the African continent, overturning decades of colonial ideology that assumed all inventions came from elsewhere (Marshall and Hildebrand 2002; Neumann 2005).

Agriculture continued to dominate the attention of environmentally minded historians, geographers, and anthropologists in the postmodern heyday of the 1990s and early 2000s, but with an increased focus on identity issues with the rise of social history. Many of these studies emphasized the cultural specificities of agricultural practices and their meanings (e.g., Fields-Black 2008; Hawthorne 2003). Several scholars deployed innovative techniques to understand aspects of history, like local perceptions of abundance and scarcity, that have often fallen outside of the historian's gaze, as well as to delve into longer-term histories of agriculture in specific regions. Historical linguistics has proved an exemplar in both respects (e.g., de Luna 2016; Ehret 2014; Schoenbrun 1998; Stephens 2016, 2018a, 2018b). Social history approaches are essential to understanding some of the more qualitative aspects of food security, particularly food preference. Narrative plays a major role in how these scholars are able to broach social history, and informs my writing style and approach in this book.

Food history is still what we might call an emergent field in Africa, despite the meteoric rise of food studies and food history in recent decades. In part, this relates to a lack of traditional source materials like recipes, cookbooks, and written archives as compared with other world regions. I suspect the paucity of attention paid to Africa is also a function of the common misconception that people struggling with food insecurity have little choice over what they are eating (e.g., Van Esterik 2006). Works like James McCann's (2009) *Stirring the Pot* and Carney and Rosomoff's (2009) *In the Shadow of Slavery* have helped put African food history on the map, and have attempted to dispel this stereotype by presenting African cooks and farmers as skilled and strategic actors, particularly in the Atlantic exchange (see also Carney 2001; La Fleur 2012). Yet in most food studies scholarship, food security is rarely addressed or critically engaged. This divide is one I try to bridge throughout this book.

Another related trend is histories of poverty, both by economic historians (see below) and by social historians (Iliffe 1987; Stephens 2016, 2018a, 2018b). While few studies in this vein directly engage food (but see Bonnecase 2018), their findings are generally applicable to understanding shifts in poverty and thus food access over time and space. Rich conceptual histories of wealth and abundance are also relevant here for outlining not only the specific manifestations of these concepts in different groups and over time (e.g., Schoenbrun 1998; Stephens 2016), but also their implications for food security. In short, the intellectual ingredients for a history of food security in Africa are already present. In the next section, I consider some of the changes in food security over time that we can deduce based on these studies. The spatial and temporal coverage of relevant studies is patchy. I want to stress that my intent is not to make a pan-African argument for changes in food security, but instead to illustrate the major shifts suggested by the available data. I argue that we need to evaluate these potential shifts in food security by building empirical histories of food security, and I conclude this chapter with a consideration of the methods for doing so.

SHIFTS IN FOOD AVAILABILITY AND ACCESS IN AFRICA'S PAST

Combined, these diverse bodies of research hint at major shifts in the productive capacity of African agriculture and document the agility of farmers over the last five millennia and more. While a continent-scale review of all of these changes is beyond the scope of this chapter, I briefly mention several shifts to foreground the Banda case study in wider context, and to provoke a series of initial questions that we might ask of a *longue durée* history of food security in Banda and more broadly.

Most archaeologists interested in African agriculture have focused on its emergence between ten thousand and three thousand years ago—a key transition in the productive capacity of many groups that has implications for food security, though it is earlier than the time frame covered in this book.[2] Although transitions like this that occurred in deep antiquity may seem to have little relevance for modern agriculture, some of the big-picture findings about this period demonstrate the importance of a *longue durée* view of agriculture and food security. For example, unlike in some world regions, where agriculture largely supplanted hunting and gathering strategies, case studies from across the continent indicate that people continued to use wild animals and plants alongside domesticated varieties for a considerable time, and that many people preferred to continue foraging as a primary subsistence strategy (e.g., de Luna 2016; Neumann 2005). Marshall and Hildebrand (2002) suggest that early domestication was motivated by a desire to reduce risk, a concern that motivates more recent farmers as well (Richards 1985). By the start of the common era, people living in more aggregated towns and cities showed various subsistence specializations, from farming to fishing to hunting (McIntosh 2005), a tendency which also reappears in later societies and probably has a very ancient origin (cf. Ehret 2014).

In considering shifts in productive capacity, we would do well to note strategies like risk reduction and specialization that reoccur in African food systems, not because their continued presence suggests that agriculture is unchanging, but because it suggests a deep reservoir of historical knowledge that can be accessed when people face problems similar to those of the past (cf. Moore and Vaughn 1994).[3] These recursive problems and solutions may be at odds with the narrow focus of many Western scientists on overproduction in the pursuit of surplus. Most classic models of agricultural origins focus on this as a universal goal (see de Luna's [2016] excellent critique of this concept in regard to African hunter-gatherers). Agricultural systems have periods of boom (harvest) and bust (just before harvest) that require storage throughout the year. For some crops and areas of the world, the ideal is to store as much as possible, in order to get people through the bust season as well as through a bad year or two of harvest. Storage also leads to the potential for accumulation of surplus by individuals or groups, who can then restrict food access for more vulnerable groups. Most attention to this dynamic has been devoted to studying the often large, well-organized agricultural systems of hierarchical societies. However, many of these same societies are also characterized by high levels of inequality, which can lead to differences in food access. Archaeologists have observed that African case studies rarely have the characteristics of a classic hierarchical state, instead suggesting more heterarchical organization, in which multiple interacting parts may or may not have power over other segments or groups (e.g., McIntosh 1999). Ensuring more equitable food access may have been politically advantageous in heterarchical systems, though we must not assume that egalitarian ethics defined political relationships for all Africans at all times (cf. Stephens 2018b, 403).

Archaeologists have also attended to the agriculture of the so-called Iron Age societies of the last two to three millennia, a time during which agriculture spread, people developed a wider array of crops and elaborated and intensified agricultural systems in sub-Saharan Africa (Neumann 2005). The success of these agricultural regimes is suggested by the emergence of the state-level societies and/or dense urban populations recorded in the Horn of Africa (e.g., Meroe and Aksum; see McCann 1999, 36–47), the Swahili Coast (LaViolette and Wynne-Jones 2018), the Great Lakes (Robertshaw 1994; Schoenbrun 1998), southern Africa (e.g., Mapungubwe and Great Zimbabwe; see Pikirayi 2002; Huffman 1996), and West Africa (e.g., Ile-Ife and Jenne-Jeno; see McIntosh and McIntosh 1981, 1984). With the exception of the Swahili coast, which came to rely on Asian rice (Walshaw 2010), each of these polities was supported by indigenous African crops like pearl millet, finger millet, sorghum, and yams, as well as livestock including cattle, sheep, and goats. Arabic chronicles of the trading empires along the Niger River provide compelling documentation for the region's surplus production of agricultural goods (Lewicki 1974; chapter 3). Where environmental records are available for comparison, it appears that some of these societies also demonstrated a high degree of resilience to climatic amelioration (McIntosh 2005). In other societies where drought seem to have had a more significant impact, social pressures seem to have also played an important causal role in their decline (Huffman 1996; Pikirayi 2006; Taylor, Robertshaw, and Marchant 2000). The emerging picture, then, is that during this period African crops and livestock enabled dense populations and a degree of resilience to environmental change, at least in the regions cited above. While these findings hint at the possibility that these societies maintained high food security, food-related data are not available for most of these areas at present (important exceptions are Murray 2005 for Jenne-Jeno and Walshaw 2010 for Swahili). Future studies might investigate whether food was distributed equally among constituents of these politically centralized societies, as well as the strategies people used to produce and access food during environmental and economic shifts.

Despite a profusion of archaeology devoted to the study of the last five hundred years, few archaeologists have focused on agriculture during this period, largely ceding the topic to historians (see Gallagher's 2016 review), though this appears to be changing in recent years (e.g., Gijanto and Walshaw 2014; Logan and Cruz 2014; Logan 2016a, 2016b; Monroe and Janzen 2014; Walshaw 2010). There is considerable need for archaeological and linguistic study of the movement and roles of plants and animals during this period. What Africans actually did and the decisions they made, especially on their farms or in their kitchens, were often very different than the views accessible to European observers. This is significant, since many disciplines rely on European and colonial records not only to understand the colonial period but also to project these dynamics back onto the precolonial past (chapters 3 and 4).

The arrival and adoption of American crops, especially maize and cassava, has received the most attention, since these crops are thought to have improved the productive capacity of African agriculture due to their short maturity time (maize) and ability to grow on poor soils with less labor (cassava). Given the presumed importance of this shift, it is surprising how few archaeological data are available on the topic (Gallagher 2016). Understanding this process is necessary for evaluating the capabilities of African agriculture prior to European interventions. As detailed in chapters 2 and 3, American crops took centuries to become staples in Banda. We find a similar hesitancy to abandon local crops in almost all cases where systematic recovery and analysis of plant remains has been pursued (Gijanto and Walshaw 2014; Esterhuysen and Hardwick 2017; Widgren et al. 2016), an important distinction since maize presence is often inferred indirectly (e.g., see review in Widgren et al. 2016). This finding is at odds with the historical literature, which focuses on coastal enclaves (e.g., Alpern 1992, 2008; La Fleur 2012; McCann 2005), where agricultural production was often geared towards the provisioning of European trade ships (Carney and Rosomoff 2009). Two recent books on the adoption of American crops on the West African coast argue for considerable African agency and ingenuity in the use and production of these new foods (Carney and Rosomoff 2009; La Fleur 2012). These are exceptional contributions to a literature dominated by Crosby-influenced interpretations of the Columbian Exchange (see chapter 2), and they provoke a series of questions about the relative roles of indigenous and introduced crops as well as about the food security of African societies during the Atlantic era.

In addition to revealing much about successful agricultural strategies in the past, findings like these are important because they can inform recent attempts by economic historians to quantify agricultural production in Africa in precolonial times. Many such studies note the poor productivity of African farms compared to those elsewhere in the world, since in these data land is generally plentiful but labor is often scarce, a ratio that favors extensification rather than intensification strategies. Coupled with environmental "limitations" like poor soils and restricted distribution of livestock, the result has been an agricultural system with low productivity per unit land (Hopkins 1973). This low productivity made it more profitable, according to some economic historians, to transport labor from the African tropics to the Americas, where agriculture was more productive (Austin 2007, 2008a). I take no issue with the general observation that lack of labor may result in lower agricultural productivity, a point I return to in chapter 4. But there are some methodological concerns with the "timeless" application of this equation. Estimates of apparently low agricultural productivity in Africa are generally derived from colonial sources, which present imperfect records of production given that centuries of slave trading had forcibly removed generations of African men and women in their primes, depriving African farms of significant amounts of labor as well as skill (see chapters 3 and 4). So too must we be cautious of classifying all of the continent's diverse ecological niches as limited; as Schoenbrun (1998) artfully illustrates, some places offered abundant resources. The archaeological examples cited above suggest that agriculture was capable of supporting large, dense urban populations, and at the very least that low productivity did not define all times and places in the continent. The question then arises as to whether Africa's environment(s) are really the limiting factor in agricultural production, or whether political and historical factors also played and continue to play strong roles.

Historical research suggests that many parts of the continent have witnessed major economic and demographic shifts over the last five hundred years, with major implications for the ability to access food. Walter Rodney's (1972) *How Europe Underdeveloped Africa* blamed Europe's extractivist endeavors, especially the trans-Atlantic slave trade, for retarding economic and demographic growth in the continent. This argument implies a serious reduction in entitlements and consequently food security. More recently, Pomeranz (2000) postulated a Great Divergence, in which the so-called developed world experienced major leaps in economic growth in the nineteenth century, while other regions such as the African continent did not. Working with what is now known as the reversal of fortune (RF) hypothesis, Acemloglu and colleagues (2002) compared the economic development of non-European areas at two points in time: 1500 (via population density and urbanization) and 1995 (via GDP). They found that the parts of the world that were most developed in 1500 (Africa, South America) were among the poorest countries in 1995, whereas the areas that were poorest in 1500 (the Americas, Australia) were among the richest countries in 1995. They explain this divergence on the establishment of European institutions like private property rights, which tended to be strongest in regions with significant settler populations. Nunn (2008) elaborated the RF thesis by a specific consideration of Africa-side dynamics including the trans-Atlantic slave trade.

The RF thesis has effectively energized a generation of economists and historians to revisit precolonial history and think about today's economic problems as results of change over the *longue durée* (e.g., see Akyeampong et al., 2014). However other economic historians have pointed out limitations of the thesis and the database on which it relies, particularly concerning the quality of the precolonial data (Austin 2008b; Hopkins 2009). Population density data are largely derived from colonial sources which are then adjusted for certain events (e.g., the slave trades) and projected back in time—a practice that experts consider highly problematic (Manning 2014). Further, Austin (2008b) notes that the decision to compare only two points in time—1500 and 1995—compresses the history of the intervening centuries. To these critiques, we might add concern about the use of patchy data to extrapolate continent-wide generalizations. Curiously, archaeologists remain largely unaware of the RF thesis despite the fact that our data sets are well equipped to address it. We regularly determine the size and density of settlements, and tend to focus on other kinds of relevant economic data on trade, for example, that would help refine or challenge some of the more problematic aspects of the RF thesis as well as provide texture and detail to these narratives at the regional level.

As with Rodney's argument, we could extend the RF thesis to postulate a major drop in food security between 1500 and the late twentieth century, since a decrease in economic well-being often results in food insecurity. I explore this argument throughout this book by examining the intervening centuries to draw out the regionally grounded long-term processes of disenfranchisement that are so far missing from the RF thesis (Green 2019). The Atlantic slave trade, which formed the central backbone of Rodney's (1972) argument for the underdevelopment of Africa, is one such process. Rodney argued that the Atlantic slave trade had resulted in stagnant population growth, especially in the eighteenth and nineteenth centuries. The implications for agriculture are clear—especially if one puts stock in the argument that labor has been the limiting factor for agricultural productivity in many parts of the continent. Carney and Rosomoff (2009) build on this and explicitly argue that the slave trade depopulated many regions of farmers in their primes, resulting in not only a labor deficit but also a brain drain. The timing proposed by these scholars appears to coincide with Banda's experience, as I explore in chapters 3 and 4.

Still, when we look outwards to other case studies, the takeaway points are different. Focusing on Central Africa (especially the Kongo kingdom), Vansina (1990, 211–16) documents increasing demand for foodstuffs by European slaving vessels on the coast, which was met by increasing agricultural production of a new cultigen, cassava, and by establishing slave villages and farms. This finding supports the conclusion drawn by Carney and Rosomoff (2009) that American crops were well suited to the trade in human beings, since their production could be scaled up effectively. But other African polities, particularly those in the hinterlands, appear to have developed different strategies. For example, the political organization of some African polities, like the Sokoto caliphate, appears to have afforded high levels of food security for some regions during the upheavals of the nineteenth century (Watts 2013), although it is unclear whether or not everyone enjoyed equal access to food supplies (probably not, as Green's [2019] argument for increasing inequality suggests). These differences prompt important questions about the relationships between agricultural production, food security, and the slave trades— and their lasting consequences across diverse landscapes, some of which I take up in chapter 3.

If our understanding of agricultural production in precolonial periods is limited to some degree by the paucity of data (Widgren 2017), we know comparatively much more about changes in agriculture during the colonial era. Increasing agricultural production was a central goal of most colonial administrations in the African continent, so there is a better archive for this period, with some important caveats. Most of the focus was on documenting the production and potential production of cash crops rather than subsistence crops. As Moore and Vaughan (1994) detail, farmers often strategically underreported their harvests or simply moved out from under the gaze of colonial officials. For this and many other reasons, Sara Berry (1993) argues that agricultural data for the African continent is woefully insufficient. This state of affairs makes it exceedingly difficult to evaluate the impact of colonial policies. Iliffe's (1987) continent-wide survey suggests that impacts were variable over time and space. One central concern revolves around the impact of cash-cropping, a strategy that most colonial administrations adopted to fund their African colonies (see Berry 1993). The cultivation of cash crops effectively siphoned land and labor away from the production of subsistence crops in the service of growing global commodities like palm oil, cocoa, and cotton (Mandala 2005; Watts 2013). In other contexts, colonial officials actively policed what they saw as wasteful agricultural techniques like slash and burn, which provoked a range of local responses (Moore and Vaughan 1994). For regions lacking cash crops, the impacts of colonial interventions are more slippery but not impossible to trace, as we will see in chapter 4.

The new science of nutrition emerged beginning in the 1930s with the realization that inadequate food intake resulted in health problems, and that these relationships could be measured. Medical doctors in the colonial service were often tasked with documenting and identifying these relationships, and their reports provide a detailed glimpse into the health of the poorest colonial subjects in particular (Worboys 1988). Height data from military conscripts has also been used by economic historians as a proxy for nutrition, since height is dictated by protein intake in addition to genetics (e.g., Austin, Baten, and van Leeuwen 2012). Yet studies relying on these data would benefit from insight as to the context of this suffering, including the economic and social positionalities of the patients.

Refining the method for understanding colonial food security is important not only because of what it reveals about food security during that period, but also because the colonial period is often used as a baseline to understand deeper, precolonial histories. Even Audrey Richards, who was interested in the changes induced by colonial policies, did not acknowledge the changes that occurred among the Bemba prior to the start of her fieldwork in the 1930s (Moore and Vaughan 1994). One of the main challenges is the source of the data itself: records created by colonial officers or scientists working for them. As Tilley (2011) has argued, colonial officials and scientists sometimes had the interests of their subjects at heart, and many documented the problems with colonial policies, particularly concerning agriculture (Moore and Vaughn 1994). However the pendulum also swung the other way, as we will see in chapter 4, when officials in the British Gold Coast actively suppressed the study of a medical doctor who found extreme levels of malnutrition in the Northern Territories, for fear it would generate negative press at home. This case demonstrates the importance of reading against the grain through using multiple archives, something that Africanist anthropologists, archaeologists, and historians have been remarkably adept at doing.

EXCAVATING AFRICAN FOOD HISTORY

In this book, I propose that we engage food security in Africa's past by focusing on its food history writ large. As McCann (2005) argues, a broader focus on food history acknowledges that food is about more than agronomic potential, and includes a range of tastes and textures produced by skilled cooks and enjoyed by people of all classes. Food history thus allows us to populate agricultural and environmental histories and to get a better understanding of food preference and desires. While many disciplines and associated methodologies have provided clues into African food history, there has been little explicit discussion of *how* we construct African food histories.

Two approaches have dominated studies of related topics in the past: the quantitative approaches of economic historians and food security analysts, and the more qualitative engagements with context of social and food historians. Economic historians prefer large, quantitative datasets that are used make comparisons between continents and time periods. They tend to test big ideas. The scale of their ideas and the influence of economics means that their conclusions spread more widely among academics and are more likely to inform policymakers, even if their application to specific settings is often inappropriate. There is a major danger in using examples from selected regions to formulate patterns about the continent as a whole, a practice that reifies the Africa-is-a-country stereotype. By contrast, historians and anthropologists largely rely on approaches that permit contextually sensitive portrayals that are more effective at investigating underlying causes and complications. However these narratives tend to be dense and specific, which limits their accessibility to scholars in other fields and to nonacademics.

I take particular insight from Fairhead and Leach's mixed-methods approach, because their study has been successful in rewriting narratives about African environmental practices and in reaching both academic and professional audiences. Like many environmental historians, they marshal empirical data drawn from environmental sciences, together with archival and ethnographic data that fleshes out cultural histories and contexts. They do so in an explicitly contrarian frame, seeking to overturn a specific stereotype about land degradation and its causes in Africa. In this book, I make use of empirical data derived from archaeology to reconstruct past food practices, as well as archival and ethnographic accounts that help add flesh to these empirical bones. I adopt a narrative framework that explicitly aims to challenge the central tenets of the scarcity slot, in an effort to produce a new narrative about food security and food history in Ghana, with implications for pursuing similar histories elsewhere in the continent.

What is also different about my approach is its coverage of the *longue durée*, from about 1400 to 2014. To capture this long time frame, I rely on the empirical databases and conceptual tools of archaeology. Archaeological data extends our timeline back almost indefinitely. As I will illustrate in subsequent chapters, scholars often read scarcity into past foodways because the data sources they consult are from much later in time, and thus from a different political and economic context than the period under study. This practice of baselining reifies the assumption that little has changed in regards to foodways, and reinforces the third tenet of the scarcity slot. Archaeologists anchor the timing of certain events, like the introduction of American crops, in chronological space that is for many regions beyond the reach of traditional documentary archives (but see La Fleur 2012 for an excellent use of documentary and linguistic sources in tracing the arrival of these crops on the Ghanaian coast). Archaeological chronologies often have ranges that are much coarser than the calendar dates of historians, since we tend to rely on radiocarbon dates that give a broader range (a date plus or minus a standard of error).

Archaeological data also reveal a different perspective on food than written archives. We study the material remains left behind as the results of people's past activities, decisions, and experiences, providing a ground-up view of everyday life (Robin 2013; Stahl 2001). The scale of this kind of history is ideal for tracing the kinds of things that McCann (2009) notes as so critical to unraveling African food history—women's knowledge, ingredients, and diversity of techniques. Historians have long made use of archaeological data to anchor certain events in time, but archaeology is usually supplementary to their main arguments (see also Robertshaw 2000; Vansina 1995) with the important exception of some historical linguists who demonstrate high proficiency in archaeology (de Luna and Fleisher 2019; Ehret 2002; Schoenbrun 1998). In this book, I make material data central because they are so well suited to revealing food history. Archaeology's ground-up perspective provides us with a unique scale from which to examine human history, one that avoids the problems of top-down history (see Stahl 2001). Material remains allow us to forefront human agency because we can trace what people actually *did*, rather than rely on what they say they did (as with historical linguistics) or what others said they did (as with most written archives) (see de Luna and Fleisher 2019).

This unique vantage provides us a sideways glance at received histories, making archaeology an especially critical toolkit for dismantling dominant assumptions like the scarcity slot. Žižek (2008) argues that in order to see pervasive structures and processes we need to observe and document them using a "sideways glance," an alternate viewpoint gleaned from different methods or starting assumptions. Food security is usually viewed from a presentist perspective, which limits the set of possibilities for the past and future and obscures the power relations at the heart of modern development. The *longue durée* view pursued in this book does important analytical work by revealing the historical processes that created present-day food insecurity but also by using the past as foil to the present. By sifting through the past, we are also able to evaluate what passes for common knowledge about African foodways.

Excavation is the primary method of field archaeology, and inspires my approach to African food history. Archaeologists excavate sites by carefully removing one layer of sediment at a time. That layer can be defined in various ways, usually by means of the natural or cultural stratigraphy (based on color and texture of the sediment), or of arbitrary levels (e.g., 10 centimeters) designated by the archaeologist. Ideally, each layer corresponds approximately to a period of time, although in practice this is often more complicated than it first appears. Nevertheless, each layer forms a basic unit of comparative analysis for archaeologists. Similarly, I divide Ghana's food history into rough periods, as presented in the next four chapters.[4] Material remains from the same level are usually collected and interpreted together to form an argument about what was going on in that period. This is very important, because it insures that objects are interpreted relative to one another and to the context they come from, avoiding the problems of baselining discussed above. Results from each layer are then compared to those above and below it to generate information on trends over time.

In this book, the act of excavation is literal, as I have just described, as well as metaphorical. Metaphorically, it is necessary to peel back the layers of assumption that have built the idea of Africa as a scarce place. In each chapter, I present one or more interpretations that have defined our understanding of foodways in that period, and attempt to evaluate each one with empirical evidence from the same period. This simple methodology provides an appropriate framework for thinking about African food history, and in particular for debunking the scarcity slot. Excavation trains our focus on information drawn from the period in question. Wherever possible, I compare data that are coterminous—from the same period—rather than rely on analogies drawn from later time periods. In chapter 4, I show that the baseline most commonly used to approach African food history is drawn from the most food-insecure period in its recent history. In order to evaluate scarcity over time, we must be careful not to assume the insecurities of one period are applied to another.

Archaeologists typically compare the content of different layers with those above and below to generate an argument about change over time. This cross-period comparison is critical to food history, because it relies on the comparison of contextually anchored narratives of one time period with those of other periods, rather than on baselining. It has often been assumed that foodways change slowly, but this is not always the case. People confronted with novelty (like American crops) or challenges (like declining household income) may make rapid changes to their diets (Macbeth and Lawry 1997). A cross-period comparison allows us to understand what came before as well as the rate and scale of changes over the *longue durée*. In some senses, this cross-period comparison is one of the most convincing parts of the RF hypotheses, yet comparing two distant chronological points (1500 and 1996) leads to the compression of history in intervening centuries (Austin 2008b). Archaeology adds more layers and in so doing makes historical processes visible.

The excavation method also provides a scaffolding for comparison that is rooted in a specific locality, answering the problems with lack of context that arise from large-scale comparisons such as the RF hypothesis. Archaeologists approach change over time by comparing layers in a multiscalar sampling universe. At the microscale, we can compare layers in an individual excavation unit of varying size (from 1x1 meter to squares of much larger sizes), which represents a small sample of the occupation of that area. Most often, we eventually scale up to the level of the archaeological site, which often approximate units like villages or towns that were culturally meaningful in the past. We also compare information among sites to arrive at regional historical trends. Archaeological units are always samples of a much larger universe. While some may see this as limiting the applicability of our results to national or global scales, in the case of food history such microhistories are important sources of alternative possibilities. In this book, I do not attempt to write a food narrative that applies to all of Africa; instead I offer a counternarrative from one small region in central Ghana that challenges the tropes of the scarcity slot. In so doing, I hope Banda can serve as a point of inspiration and comparison for the construction of other counternarratives on the continent. Where possible, I compare Banda to other parts of Ghana, which brings out some of the divergent responses to Atlantic trade and colonialism.

Excavation also relies on the careful sifting and collecting of material remains from their sedimentary contexts, and an acknowledgement of the affordances and taphonomic histories of those artifacts.[5] It is impossible to construct a perfect history; we are bound by our archives as well as by the perspectives those archives communicate and silence (Trouillot 1995). In some cases, entire sets of activities or people may remain invisible. For example, in written records, women and their activities are often portrayed only in ephemeral ways, since most European chroniclers were men. Archaeology helps provide a solution, since material remains record the everyday activities of most people. But some activities have few surviving material traces; in terms of food, for example, we are hard pressed to find traces of ancient tuber crops like yams and cassava. Historical linguistics would greatly help in tracing these crops, as it has in other places in the continent, but is unavailable for Banda and surrounding areas. Instead, we are left to infer their use based on other kinds of material remains or on later records of their use. In these instances it is necessary not only to be clear about the source of this information, so that later scholars can offer critique if needed, but also to note that these arguments are less strong than those for which we have good material evidence. This kind of approach is very important for making sure we do not recreate new stereotypes while attempting to debunk the scarcity slot.

One of my primary sources of data about ancient foodways comes from archaeological plant remains. Archaeobotanists or paleoethnobotanists study three kinds of plant remains (macroremains, phytoliths, and starch grains) that come to be deposited and preserved in the archaeological record (Pearsall 2015). I rely mostly on macroremains—seeds, nut shells, and other plant parts—that are usually preserved in charred form. This means they must come into contact with fire in order to last long enough for archaeologists to recover them. Like any source of data, macrobotanical remains are subject to several preservation biases, but luckily cooking and processing activities are well represented in the remains (Hastorf 2017). Unfortunately, soft plant parts, especially tubers, are underrepresented, which limits my interpretations at times. I have tried to uncover these plants as well as activities that do not involve fire by also analyzing phytoliths.[6] Phytoliths are distinctively shaped silica casts of plant cells or intracellular spaces that allow archaeobotanists to identify specific parts of a plant (leaf, seed, glume, etc.) as well as different plant taxa (Piperno 2006). Unlike better-researched areas of the world, where methodologies have been developed to identify specific plants, analyses are only in their infancy in the African continent and as such their applications are limited as of yet (see appendix A; Logan 2012, 82–116; Ball et al. 2016). When possible, I use phytolith analysis to flesh out how different grain crops were used, but future work will surely unleash a plethora of insights that are unavailable to us at present.

Archaeobotanical remains reveal the plant-based component of past diets, which forms most of what people eat; they are one of the most reliable means of accessing this part of the culinary past. However, in order to access their full interpretive value and flesh out political and cultural contexts we must compare plant remains with multiple other kinds of data. This present study is only possible because of over thirty-five years of sustained archaeological work at Banda, under the direction of Ann Stahl. Stahl and colleagues have documented the many other kinds of material remains that archaeologists encounter and that are important for constructing food history. These include ceramics, the vessels in which food is cooked, stored, and served; animal bones, the leftovers of meat and animal consumption; metals, ranging from everyday agricultural implements like iron hoes to specially fashioned copper alloy ritual objects; and a wide array of more rare items, like ivory and beads, which attest to vibrant and diverse local economies and ritual ecologies. These data are derived from archaeological excavations at villages that span the last one thousand years, and include intensive sampling of four major archaeological sites as well as regional sampling of many more, details of which can be found in Ann Stahl's comprehensive publications (e.g., Stahl 1999b, 2001, 2007) and are discussed in each chapter. Full methodological and sampling details can be found in appendices A and B.

In chapters 2, 3, and 4, I put archaeological data directly into conversation with some of the primary arguments about the corresponding period. The scholars making these arguments come from a variety of disciplines, including history and geography, and some did their work long ago and in different intellectual climates. In most cases, these researchers had very little data on which to base their arguments, and so made logical leaps that were unfortunately based more on prevailing ideas about the African continent than on empirical data. I have selected these particular arguments not because they make easy straw men, but because they remain remarkably tenacious in how we think about African foodways. It is essential to critique each of these arguments head-on in order to change dominant narratives about African foodways and African history in general. Archaeological data not only provide an empirical test of these arguments, but also offer alternative narratives about the period in question.

For later time periods, particularly the nineteenth and twentieth centuries, I supplement archaeological information with historical and ethnographic archives, which add considerable texture and connect the material past to lived realities of the present. These chapters are critical in connecting past and present, a project that is central to the goals of this book. Yet I also acknowledge my limitations as a historian or cultural anthropologist. Interdisciplinary work requires that we go beyond our theoretical and methodological comfort zones, but one's strengths and weaknesses are bound to show in the cracks of arguments left unexplored and sources left unturned. I consulted archival sources at Ghana's National Archives in Accra as well as regional archival offices in Sunyani and Kumasi (appendix A). Secondary source material from the careful work of historians and archaeologists provides a check on the work I present here, and helps flesh out though not wholly eliminate the blind spots in my own archival work.

The ethnographic component of this work is captured in chapters 5 and 6, which focus on the last half-century or so. Most of this ethnographic study was conducted over a six-month period in 2009, followed by shorter six- to eight-week visits in 2011 and 2014. My primary goal was to understand the topography of food changes over the last few generations. I relied on semi-structured interviews supplemented by participant observation. In total, I was able to conduct 120 interviews with women and men spread over five villages with the assistance of Enoch Mensah, an exceptional research assistant and translator from Banda.[7] My focus was on women, since they are the primary cooks. A smaller number of men were interviewed regarding shifts in agricultural practices. While these data are sufficient to inform chapters 5 and 6, the relatively short window in which they were collected precludes a more comprehensive monograph-length treatment. Consequently, the reader will see that these chapters focus quite narrowly on food and women's work over time. I acknowledge that the changes and continuities I observed were part of a much more complicated cultural tapestry that I was able to cover only superficially.

Whatever my limitations as an ethnographer or historian, I find value in being pulled into the present from the past. The narrative that results is what we might call an archaeological ethnography of food, in a similar vein as Lynn Meskell's (2011) archaeological ethnography focused on South Africa. My goals are to connect the past to the present in a meaningful way, and to use those connections to speak to possible futures. To me, this is the power of the archaeological ethnography genre. This genre also acknowledges the tremendous leaps that have been made by ethnoarchaeologists in the African continent, many of whom have transcended traditional ethnoarchaeological questions to ask more culturally appropriate and ethnographically informed questions and have become advocates for the communities in which they work (e.g., González-Ruibal 2008, 2014; Schmidt and Pikirayi 2016). I also take insights from the many attempts of Africanist archaeologists to make the past useful in the present (e.g., Lane 2015; MacEachern 2018; Stump 2010). This archaeological ethnography of food also follows in the footsteps of the work of many Africanist historical anthropologists who have combined multiple archives in attempts to provincialize hegemonic discourse about Africa's past (Comaroff and Comaroff 1991, 1992, 1997; Stahl 2001; Vansina 1990). In particular, these works have attended to the critical question of how knowledge about the past is produced, which is essential for unraveling food histories as well. My goal here is to make a similar intervention in our understanding of Africa's foodways past and present.

Writing an archaeological ethnography of food makes it necessary to communicate things a bit differently, and readers will find that this volume is neither a traditional archaeological monograph nor an ethnographic or historical one. One of my central goals is to make archaeological data accessible to the nonspecialist, because this is essential for advancing a new kind of African food history. I do not dwell on the limitations of archaeological data. No discipline has access to a perfect data set. While I am cautious in my interpretations, I avoid listing or evaluating the multiple sets of alternative hypotheses that may explain certain patterns in the main text. These are important, even critical, exercises, but have already been accomplished in other published work on Banda, most notably in the work of Ann Stahl as well as my previous publications. I have also removed the customary tables of material remains that tend to be found in archaeological works from the main text and instead have focused on offering qualitative descriptions. These data as well as the methods used to produce them are available in the appendices. Abandoning these two archaeological writing conventions means I am able to focus less on objects and data and more on human experience. Following the lead of Hegmon and colleagues (2016), wherever possible I make people, rather than material types, the subject of my argument. While the archaeological portions of this book do not completely match the ethnographic ones in tone, I do my best to make both speak to everyday life in the period of focus.

Choosing Local over Global during the Columbian Exchange

The diverse African societies and landscapes that existed before European contact were very different indeed than the ones we see today. Rich with natural resources that peppered ecological zones ranging from the wettest of forests to the driest of deserts, West Africa hummed with long-distance trade networks that brought all manner of goods to a wide array of village and urban consumers. Vibrant subregional trade propelled African goods to global marketplaces, supplying villagers with prestige goods from far afield and luring Europeans to its shores beginning in the fifteenth century (Green 2019; Mitchell 2005). Over the next few centuries, West Africa was to change dramatically, a story I continue in the following chapters. On the eve of these transformations, what was life like for villagers involved in long-distance trade? How did these experiences differ among villages and social groups? Was their everyday experience one of scarcity, plenty, or something in between? How did this change as Europeans landed on the coast, ushering in one of the largest-scale global exchanges in history?

Foodways of the early Atlantic period (c. 1400–1650) have mostly been considered under the rubric of the Columbian Exchange, and are illustrative of the ways in which scarcities of the present are assumed to have also characterized the past. American crops like maize and cassava are staple foods across sub-Saharan Africa today, and their presence alleviates food security concerns for many. Because of the important role of these crops in food security today, many scholars have surmised that they were adopted to meet such needs in the past. The context into which these new foods were adopted is rarely considered. As we know from materiality studies (Ogundiran 2002; Stahl 2002; Thomas 1991), understanding context is central to accessing how new goods were received. More than that, bringing local contexts to the forefront provincializes Eurocentric perspectives that would otherwise privilege the role of European explorers in spreading these new crops (Carney and Rosomoff 2009;

La Fleur 2012). Considering local tastes and contexts helps us understand the choices that people made, and helps us evaluate whether scarcity defined peoples' experiences of food. To these ends, in this chapter I consider three interrelated questions: Were people desperate for the higher yields of maize or the dependable harvests of cassava? What were the environmental conditions at the time of the adoption of these crops? And how did preexisting food preferences shape peoples' responses to their arrival?

The arrival of Europeans also marks the beginning of our written archives for regions in the savanna and southwards to the coast (Arab records from the interior begin much earlier, starting in the ninth century AD), and most considerations of the adoption of American foods have relied heavily on these sources (e.g., Alpern 1992, 2008; La Fleur 2012; McCann 2005). Most historical sources have focused on the coastal entrepôts where European and African traders lived in close proximity rather than on the vast majority of territory outside of these zones of direct interaction. Because these sources were penned by European men, they provide only a partial view of *how* these new foods were adopted by the local cooks and farmers. In this chapter, I offer a different view of the Columbian Exchange, one that focuses on countering certain stereotypes of Africa as a scarce place. Through the archaeological and food remains at two villages, I reveal how Banda weathered a severe drought through a strong economy that afforded people access to the local foods they preferred.

BEYOND CROSBY: THE COLUMBIAN EXCHANGE

The importance of American foods in Africa is more obvious than in any other continent of the Old World, for in no other continent, except the Americas themselves, is so great a proportion of the population dependent on American foods. Very few of man's cultivated plants originated in Africa . . . and so Africa has had to import its chief food plants from Asia and America.

—ALFRED CROSBY, *THE COLUMBIAN EXCHANGE*, 2003

The potential genetic resources for agriculture in Africa were also unbalanced. Of large-seeded grass species . . . that were potentially domesticatable cereal crops . . . Africa had only four, none of which would be one of the world's primary grains in the 20th century . . . therefore, Africa had to overcome an early liability, which, of course it eventually did by adopting exotic crops . . .

—JAMES MCCANN, *MAIZE AND GRACE*, 2005

The Columbian Exchange refers to the biological and cultural exchanges that occurred in the centuries after Columbus mistakenly landed on the shores of the Americas. Popularized by Alfred Crosby's 1972 book, which was revised in 2003, the exchange is understood to have operated on a scale and with consequences that make previous world systems seem minor by comparison. Animals, plants, people, diseases, commodities, and knowledge passed between the previously isolated Western and Eastern hemispheres. Illustrative of the impact on present foodways are the facts that before the Columbian Exchange there were no tomatoes in Italy or potatoes in Ireland; both crops were domesticated in the Americas and found new audiences and new cultural significance once they were adopted in Europe.

Africa's contributions to and benefits from the Columbian Exchange are less well known. Though Crosby devotes only three speculative pages to the African continent, his interpretation has proved remarkably tenacious, as seen in the juxtaposition of quotes above. McCann (2005) also draws on Jared Diamond's (1999) environmental determinist argument, which like many explanations of crop adoption in Africa accord environments more agency than the humans who fashioned them (La Fleur 2012). This is in part simply a limitation of archive: we know much more about the properties of different crops and environments than we do about the people who decided to adopt and modify them centuries ago. Unfortunately, when information is especially limited, reasoning based on the scarcity slot tends to fill in the gaps. Both quotes referenced above imply that Africans were incapable of developing their own crops (scarcity slot tenet 2), and consequently, lacked sufficient calories to feed their populations (scarcity slot tenet 1).

Yet, even by the time of Crosby's first writing, archaeologists had shown that Africans had indeed developed quite an impressive array of domesticated plants and animals thousands of years prior to the Columbian Exchange, including pearl millet and sorghum (deWet and Harlan 1971); we now know the range of adaptations were even more impressive (Marshall and Hildebrand 2002; Neumann 2005). McCann (2005, 40) underplays indigenous crops by noting they are not of significance to global markets, but this is more of a statement of their political value than of their capabilities.[1] The most valued foods, on a global scale, tend to originate from regions of power, therefore the marginal role of African crops is not surprising given the uneven geopolitical landscape of the twenty-first century. In this chapter, I illustrate that the African domesticates pearl millet and sorghum were essential in the great civilizations that preceded European interventions on the continent and remained so until very recently. Understanding the role of African crops prior to these interventions is critical in situating their potential contributions in the future (chapter 6).

Crosby's reasoning betrays a second generalization: that African crops were not capable of meeting food security needs, or at very minimum, did not produce a meaningful surplus (see also Goody 1977; Goody 1982, 58–60). Echoes of this kind of reasoning can be found in McCann (2005) and Wilks (2004), who credit the rise of the Asante state to the introduction of maize and its ability to produce a large surplus (chapter 3). As de Luna details (2016, 8), this obsession with surplus comes from a worldview that celebrates capitalism and mercantilism, and, in a classic Othering move, equates lack of surplus with savagery and, I would argue, scarcity. Yet surplus is also the product of a certain kind of agriculture (monocropping) to fulfill a certain economic goal (export). Other kinds of agriculture may well be better suited to different economic goals (e.g., risk reduction) that do not rely on agricultural products as major exports (see the introduction, chapter 5). In this chapter, I evaluate the economic and environmental contexts of crop adoption during the Columbian Exchange to determine food security levels and thus evaluate the need for surplus.

If Crosby's portrayals of African crop adoption are inaccurate at best, what alternate models are available? While heavily influenced by Crosby, McCann's (2005) *Maize and Grace* provides an ambitious, continent-wide view of how people in diverse situations across the continent adopted maize. Three important volumes push back at Crosby's main arguments: Judith Carney's (2002) *Black Rice: The African Origins of Rice Cultivation in the Americas*, Carney and Nicholas Rosomoff's (2009) *In the Shadow of Slavery: Africa's Botanical Legacy in the Atlantic World*, and James La Fleur's (2012) *Fusion Foodways of Africa's Gold Coast in the Atlantic Era*. These authors emphasize the role of African agency in deciding whether or not to adopt new crops. This is an elegantly simple but important point, one which had long been made in the case of European adoptions of American crops: in all but the most desperate of circumstances, people have some decision-making power over what they eat (de Waal 1989). As my starting point for this chapter, I assume that the vast majority of people maintain some degree of choice over what they ingest. In the next section, I begin by highlighting the pros and cons of new and local grains that would have informed peoples' decisions. I review historical evidence of the use of existing foods and adoption of new ones on the Ghanaian coast. I then move to Banda to understand the role of local versus global crops during the mid-second millennium AD.

LOCAL VERSUS GLOBAL GRAINS

Worldwide, more maize is grown today than any other crop, and its popularity extends to the African continent. Originally domesticated in the American tropics, maize has tropical adaptations that make it ideal for tropical parts of West Africa and southern Africa (La Fleur 2012, 4; McCann 2005; Miracle 1966). Given the global importance of maize, it is not surprising that it has also received the most scholarly attention in Africa, as evidenced by McCann's (2005) recent book. If we were to rely on maize's botanical qualities alone, we might agree that it is, as McCann's title, *Maize and Grace*, suggests, the "grace" of Africa, enabling the continent to persist despite the hardships of the last several centuries. To McCann and Crosby, maize's success is due to its potential to produce high yields, and the short time required for the plant to reach maturity. But we must be mindful of the currency used to assess maize's agronomic value. Although two crops of maize can be produced per year, this double production schedule means that labor costs are also doubled, and labor has often been the limiting factor in African agricultural production (Hopkins 1973). While maize is capable of producing more calories per hectare than many other grain crops, people do not always choose the most economical choice. Instead, local taste preferences and food security play major roles in whether or not people adopt new crops. In emphasizing maize and other foreign crops, many researchers have underappreciated the rich capabilities of African grains and their potential to support large populations. In this section, I review the costs and benefits of maize alongside those of pearl millet and sorghum, two African crops that were cultivated for millennia prior to maize's introduction. This kind of comparison allows us to understand the tradeoffs that informed farmers' and cooks' decisions whether or not to adopt maize. This discussion will, in turn, be used to help understand why people in Banda adopted maize at the rate and to the extent that they did.

Pearl millet (*Pennisetum glaucum*) was the first domesticated plant in the continent, and has been cultivated in Ghana for over three thousand years (D'Andrea, Klee, and Casey 2001). It remains an important staple in the northern half of the country. Pearl millet is the sixth-most important crop on a worldwide scale, and the third most important in the African continent (National Research Council 1996, 77, 80). The grain is extremely drought tolerant and enables agriculture in even the most depauperate conditions across Africa and India, though it is less cultivated than maize because it is comparatively low-yielding and has not received nearly as much research support despite being a risk-averse choice for farmers (National Research Council 1996, 79, 97). It is nutritious, containing both a fair amount of protein for a grain crop (9–21%) and more oil than maize. Pearl millet is also versatile: it can be steamed and eaten, used to make porridges and beer (National Research Council 1996, 81), and even consumed as uncooked flour (chapter 5). The late-maturing (140–90 day) variety is widely grown throughout the interior savanna; early maturing (80–90 day) millet is found in the far northern reaches of the country, where it is the first crop of the season (rather than maize as in the south). Pearl millet generally grows in areas with 250–800 millimeters of yearly rainfall (Brunken, de Wet, and Harlan 1977, 163), but in Ghana tends to be cultivated in moister locales with an average rainfall per year of 1,000 millimeters.

Sorghum or guinea corn (*Sorghum bicolor*) is an African domesticate that is a principal cereal in the interior savanna where, like pearl millet, it is produced primarily for domestic use and local sale. Sorghum is usually planted on more fertile land, being interplanted as a late crop with maize (or, in the north, with early millet); its yields on poor land are moderate. There are a few quick-maturing varieties (110 days), but most are longer-maturing, requiring 140–90 days. Compared to pearl millet and maize, sorghum is much more tolerant of wet conditions (Staff Division of Agriculture 1962, 370) and even of waterlogging (House 1995). It is not surprising that in Banda in recent years, as rainfall has been unpredictable and thus damaging to both maize and pearl millet, sorghum has emerged as the most dependable yielder (chapter 5). Sorghum is widely grown in areas that receive

1,000 millimeters or more of annual rain, and is best suited to areas reaching 80–1,400 millimeters of rain a year, but can grow with as little as 254 millimeters or as much as 3,050 millimeters (De Wet and Harlan 1971, 130–31). If grown together, pearl millet and sorghum complement one another nicely, as millet does well even in the driest of conditions, and sorghum in situations that are too wet for millet.

How do the costs and benefits of sorghum and pearl millet compare to those of maize? Pearl millet and sorghum are overall more nutritious than maize. They also fare better in storage, since maize's high moisture content increases storage loss. Maize is also prone to insect and pest damage (Forsyth 1962, 394–96). Because pearl millet and sorghum have long coevolved with local pests, these crops may be better suited to pest management (although improved maize varieties may have some of the same qualities). Maize is highly demanding of soil fertility; over the long-term, significant investment in maize production could have negative consequences for land fertility. Finally, one of the more concerning long-term tradeoffs of maize adoption is its susceptibility to drought at key points in its life cycle, particularly during tasseling. Indigenous crops like pearl millet and sorghum, on the other hand, are well known for their drought resilience, meaning that these crops tend to produce higher yields in savanna environments (Miracle 1966, 208). In terms of labor, estimates vary significantly based on environmental factors, with Miracle's (1966, 207–13) comparisons generally seeing maize and sorghum as similar in terms of labor requirements and pearl millet as more costly. The initial processing of maize, which involves removal of the kernels from the cob, is easier than the same stage for pearl millet or sorghum, decreasing requisite time and labor costs. These savings may be lost in later stages of processing, particularly grinding or pounding into flour, which Banda women told me is more laborious for maize than for pearl millet.

Comparing yields of maize to sorghum and pearl millet is complicated by environmental factors as well as labor availability. In Ghana in recent decades, maize is a high-yielding crop par excellence, particularly with the application of fertilizers, yielding 350–800 pounds per acre. Native pearl millet (200–600 lb./acre) and sorghum (300–700 lb./acre) yield consistently less, and respond less well to fertilizer (Staff Division of Agriculture 1962, 369–72). However, synthetic fertilizers would not have been available at the time of maize's introduction several centuries ago. Miracle (1966, 207–8) suggests that maize is actually a comparatively low yielder in savanna and forest margin environments because of its proneness to drought. And while maize lends itself more readily to intensification, labor costs may have been a limiting factor in many African situations (Hopkins 1973). Sorghum and pearl millet make much more sense when people choose extensification strategies.

Maize's most significant advantage is its ability to produce a crop more quickly (three to four months) relative to pearl millet and sorghum (five to six months; but see below regarding early-maturing millet). In tropical West Africa, this short maturity time lines up perfectly with the two-peak rainfall pattern of equatorial regions, so that two harvests can be grown per year rather than just one, potentially doubling yields. In terms of food security, the timing of maize's maturity is critical:

it is ready to harvest as early as July or August, which falls during a period of food shortage known as the hungry season gap (McCann 2005; Miracle 1966). The hungry season gap is created in part by the maturity schedule of indigenous cultigens like yams, which in wooded savanna regions like Banda are not ready until August or September, while sorghum and millet are usually harvested in November and December (chapter 5). The first crop of maize is ready precisely when people need it most, in July, when previous grain stores are running low prior to the maturity of other indigenous staples. Considering all of these factors, it seems likely that maize may have been adopted as a stopgap crop to prevent seasonal food insecurity or as a way to intensify agriculture in situations where labor was available.

While it is tempting to rely on the botanical qualities of maize as compared to those of local grains to explain maize's adoption, agriculture is a social and environmental technology that is intricately related to the political and economic contexts in which it operates (Guyer 1984, 1988). While several excellent reviews chronicle the existing data on American crop adoption in West Africa (Alpern 1992, 2008; Gallagher 2016), few works take full stock of the political, economic, and environmental contexts into which these crops were introduced. This is in part the shortcoming of the review article genre, but it also expresses a limitation of archival sources. An unintended consequence of this limited archive is that such adoptions appear to have taken place within a blank cultural canvas. The few archaeological studies that do exist mostly date to periods that were centuries after maize's initial introduction (Gijanto and Walshaw 2014; Kelly 1995; Maggs 1982; Norman 2009).

FOODWAYS IN COASTAL GHANA IN THE EARLY ATLANTIC ERA

Coastal Ghana in the late sixteenth and early seventeenth centuries is one of the few places in Africa where written historical sources allow the reconstruction of foodscapes in the decades immediately after the introduction of some Atlantic crops, particularly maize. One excellent primary source (de Marees 1602 [1987]) and La Fleur's (2012) recent book on fusion foodways together provide a relatively high-resolution picture of foodways based on historical and linguistic data. These sources help re-create a partial image of coastal foodscapes that provides a broader regional foodscape for Banda, and they demonstrate how new crops played variegated roles across time and segments of society. While many new crops were introduced during the first century or two of European encounters (e.g., plantain, sweet potato, etc.; Alpern 1992; 2008; Gallagher 2016; La Fleur 2012), I focus on maize, since this crop sheds unique light on food security and food choice in the early Atlantic period.

Dutchmen Pieter de Marees visited coastal Ghana in 1601, and his relatively detailed description of foodways and agriculture provides one of the best primary sources for this period. In multiple places, he mentions that people along the coast had a sufficient food supply (de Marees 1987, 41), a point omitted by the historians who rely on this source. Yet he also details that not everyone enjoyed the same access to preferred foods. The elites seem to have had primary access to particular kinds of meat (chicken, goat, ox, and venison) (42). Yams are mentioned as the most common food of Africans (164). This starch-rich tuber would have provided more than ample carbohydrates (contra McCann's [1999, 119] suggestion that the forest was lacking in this regard; see chapter 3). Other plant foods noted by de Marees included *millie* (likely pearl millet), rice (probably African *Oryza glaberrima*), bananas, beans, palm oil, and possibly sweet potatoes (40, 159). Spices included ginger and grain of paradise (*Aframomum melegueta*), a spice grown only along the coast and highly sought after in medieval Europe (160).

In terms of daily foodscapes, women were apparently responsible for both procuring and making food. Each day they threshed and ground only enough millet for the day's meals (40). Men grew the crops, working first on the headman's fields before their own, for which they were repaid in food, drink, and merriment (111). De Marees (110) describes their agricultural cycle: planting *millie* and maize after a period of storms followed by heat, most likely at the onset of the rainy season (the translators note that this is after the transitional period between dry and wet seasons). The millet, he reports, "grows and flowers within three months, after which it is cut and left on the field for another month in order to dry" (112), which seems to indicate early-maturing millet. This is important, for it suggests early-maturing varieties were on the coast, and would have effectively prevented a hungry season gap from an agronomic perspective. If so, there was likely no gap that maize needed to fill, and here it is notable that indeed de Marees mentions no such shortage. Lacking a hungry season, maize is a solution in need of a problem. What seems probable is that maize was cultivated in the same schedule as early maturing millet, which it may have eventually come to displace.

De Marees devotes a chapter to the introduction and adoption of maize, which would have been a novel crop to both Europeans and Africans at the turn of the seventeenth century. He mentions that pearl millet was the grain Africans always had grown, and that they made do with it prior to the Portuguese arrival, when maize was introduced. By the time of his arrival, maize grew in abundance, and some Africans ground it with pearl millet, sometimes eating a half-and-half mixture, which was apparently prepared into a bread (113).[2] Maize produced two crops per year. He mentions both small and large maize growing, of colors including white, black, yellow, purple, and more. It is unclear what he meant by large and small maize; La Fleur (2012, 93) interprets this as representative of multiple maize varieties, but it seems likely to the translators (114n5) and to me that at least one referred to sorghum, which is otherwise conspicuously absent from his account. Sorghum comes in a range of colors, including the hues described by de Marees. This diversity is more likely for sorghum, which had long been cultivated in West Africa, than for maize, which due to its recent and limited introduction was probably subject to a severe genetic bottleneck. Maize and sorghum plants look practically identical until flowering, so it is highly likely that these two were confused by de Marees, who was not a botanist (the same oversight may well plague other chroniclers like Bowdich [(1819) 1873]; see chapter 3).

La Fleur (2012, 55–61) suggests that it was the Portuguese, not local Africans, who were plagued by food scarcity, particularly in the early days of their settlement in the mid- to late fifteenth century. Supplies of familiar foods—wheat, wine, and olive oil—from Europe were not dependable, making the fledging coastal fort of Elmina completely dependent on African farmers. The Portuguese were, however, highly suspicious of unfamiliar foods, and avoided them when possible. Imported, familiar foods were likely claimed by the fort's elite leadership, while the Portuguese of low rank as well as the Africans they had enslaved most likely relied on local produce. That preferred foods were often unavailable meets two of the three criteria for at least periodic food insecurity, and may have been perceived as quite the hardship despite the ready availability of African foods. In contrast, local African elites were fascinated by new foods, often trading items of high value to gain access to culinary novelties (La Fleur 2012, 63). This contradicts modern understandings of African tastes as tied to quantity rather than novelty (de Garine 1997), suggesting a very different socioeconomic setting. While the details of these encounters and adoptions remains shrouded in mystery due to the limited nature of historical archives for this early period, what we do know suggests a significant inversion from traditional historiography of Africa during the Columbian Exchange. In this instance, it was Europeans rather than local Africans who experienced food scarcity and were in need of carbohydrates.

The written record helps us understand some of the brushstrokes on the cultural canvas into which maize arrived. The adoption of new crops was clearly structured by existing and emergent socioeconomic inequalities, as well as existing food preferences, culinary practices, and agricultural techniques (La Fleur 2012). The canvas can be brought to life by investigating what people did in the villages and households into which new crops, particularly maize, were introduced. Examination of plant remains from archaeological sites provides one of the few ways to understand how maize was adopted at a household level. Only recently have archaeologists began to address this issue directly through examining the crops remains themselves as well as their economic, social, and environmental contexts.

FOODWAYS AND POLITICAL ECONOMY
IN BANDA, AD 1400–1650

In Banda, much of our data on daily life before and during the Columbian Exchange comes from two towns we know by their archaeological site names, Ngre Kataa and Kuulo Kataa, named after their proximity to modern-day villages.[3] These are among the largest villages occupied during the early Atlantic era in the Banda area, which we know archaeologically as the mid- to late Kuulo phase (AD

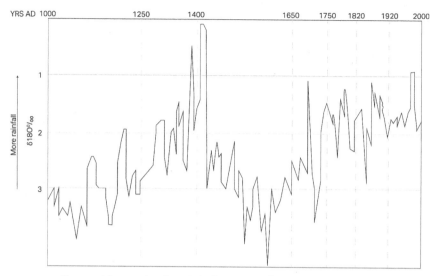

FIGURE 1. Oxygen 18 isotope data from Lake Bosumtwi, Ghana as a proxy for precipitation (based on Shanahan et al. 2009, 379). Period from 1400 to 1650 corresponds to time covered in chapter 2.

1414–1615), based on distinctive pottery forms and radiocarbon dates (Logan and Stahl 2017; Stahl 1999b, 2001, 2007).[4] Both sites were occupied over several centuries, with people's repeated activities resulting in contexts accumulating on top of one another, forming low mounds across today's landscape.[5] Multiple mounds were tested at each site in multiple excavation series in the hopes of uncovering domestic, craft working, and midden areas (Stahl 1999b, 2001). Excavating these different kinds of spaces allows us to reconstruct the political economy at Banda at the height of trans-Saharan trade and the later shift to the Atlantic trade.

The early Atlantic era coincides with the worst drought to hit central Ghana in centuries. Paleoenvironmental data from Lake Bosumtwi, located about 200 kilometers south of Banda, indicate markedly arid conditions from about 1400 to 1650 AD, with a particularly pronounced spike of aridity around 1400 (figure 1; Shanahan et al. 2009). If modern responses to drought (chapter 5) are any indication, we would have expected this far more severe, prolonged drought to have been catastrophic for daily life. In the paragraphs that follow, I attempt to recreate what life might have looked like in this very different version of Banda, following Hegmon and colleagues' (2016) writing conventions, which aim to highlight human experience rather than material remains. For more details on the archaeology of Banda during the mid- to late Kuulo phase, see Stahl's extensive publications (Stahl 1999b, 2001, 2002, 2007, 2015, 2018a, 2018b) as well as a more recent consideration of the relationship between environmental change and exploitation of plants and animals (Logan and Stahl 2017). Data tables are available in appendix B.

While we do not yet know the extent of Ngre Kataa, nearby Kuulo Kataa was one of the largest, densest villages in the Banda area at about twenty-eight hectares or the size of fifty-two American football fields (Stahl 1999b, 16). Probably the location of a large regional market (Stahl 2018a), Kuulo Kataa must have been a vibrant place indeed during the height of the trans-Saharan and early Atlantic trade. If we were to walk through the village, we would have seen a wide array of people practicing different crafts. Skilled potters fashioned all of the storage, cooking, and eating vessels one could imagine, with distinctive grooved and stamped decorations along their rims and shoulders (Stahl 1999b). Specialized potters fashioned some of these pots in the village itself, while other shapes and sizes were likely obtained through regional trade mostly east and sometimes west of the hills (Stahl et al. 2008), perhaps in exchange for foodstuffs, as recorded in much later accounts. Women most likely made the pots, if ethnographically known trends held true (Stahl and Cruz 1998). Potters retained a relationship with iron workers, presumably male, attested by the presence of slag temper in vessels (Stahl 2016). Metalworking was extensive at Kuulo Kataa and Ngre Kataa, where men fashioned both iron and imported copper alloys into utilitarian tools like iron hoes, as well as fancy objects in the shape of seated, human-like figures and others mimetic of snakes and used in ritual practices (Stahl 2013, 2015). Rituals were one way that people made sense of and tried to influence events around them, like the rains, as well as to attain other personal and political goals (Logan and Stahl 2017, 1394–95; Stahl 2008, 2015, 2018b). In some households at Kuulo Kataa, people were busy working hippo and elephant ivory for trade with outsiders, filling the high demand for this material at home and abroad (Stahl and Stahl 2004). Locally produced ivory goods may have been exchanged for items of value, such as glass beads from trans-Saharan networks. If recent ethnographic depictions held true, imported beads may have been used in women's nubility rites, thus "inscrib[ing] subcontinental exchange on local bodies" (Stahl 1999b, 37–38).

We cannot yet discriminate many differences in daily life at neighboring Ngre Kataa, except that the village seems to have been smaller and to have lacked a large regional market. Some of these differences may be attributed to the earlier initial occupation of the village, which began in the previous, Ngre phase, from AD 1230 to 1400 (Logan and Stahl 2017, 1361). Potters, metalworkers, and farmers inhabited Ngre Kataa as well, though we have less evidence that people were crafting ivory. Ngre Kataa may have been known for the skilled metalworkers who lived there, as evidence attests to a range of activities including lost wax brass casting (Stahl 2013, 2015). The diverse array of craftspeople present in the Banda area was probably critical to creating and maintaining the region's wealth. In some parts of precolonial Africa, wealth was not measured by the accumulation of goods, but by the accumulation of people with a range of different skill sets (Guyer and Belinga 1995; Richard 2017). While we cannot be sure that this value system applied to Banda specifically, this economic system appears to have been associated with a high degree of resilience to environmental change, as I argue below.

Peoples' relationships with animals were quite extensive, and meat was consumed in a quantity unmatched earlier or later in time (Logan and Stahl 2017; Stahl

1999b). The casual visitor would have likely seen plenty of sheep and goats roaming the villages, along with cattle, a great variety of birds, including chickens and guinea fowls, fish, and the occasional pig.[6] Larger domesticates appear to have been butchered into many smaller pieces, perhaps to provide meat to a larger number of people or to facilitate preparation in stews or sauces. But alongside these more common animals were a host of strange and exciting species with colorful pelts and fierce teeth and claws, a veritable menagerie whose origins spanned from the dry savanna to the dense tropical forest (Logan and Stahl 2017; Stahl 1999b). Not only would these animals have provided a range of meaty textures and flavors, from gamey and musky to fatty, but they would have supplied a range of valuable pelts and paraphernalia for chiefs. Highly skilled hunters must have acquired the most dangerous animals, including hyenas, lions, hippos, leopards, and warthogs (Stahl 1999b, 33). Curiously, peoples' tastes were not satisfied with the animals from the immediate vicinity alone. Many species were acquired from tropical forests some distance away (e.g., various monkeys; Stahl 1999b, 33). A few rare exotics must have come from the coast, including great white sharks (Logan and Stahl 2017), and may have served nondietary purposes such as personal adornment.

Outside of the villages people grew crops which formed the mainstay of their diets. We cannot yet reconstruct the specific range of cultivation techniques used, or how they might differ from those in use today, but archaeobotanical evidence allows us to paint the broad contours of what was grown and how. Unlike in the more arid regions to the north, there is little evidence for monocropping of grains. Instead, people probably practiced shifting cultivation techniques, with fields containing a mix of grains, tubers, beans, and more. We know this because we see so few weedy species mixed in with domesticated crops, which suggests that people harvested those domesticates individually, by hand,[7]

As discussed more fully below, pearl millet appears to have been the staple grain of choice (see appendix B). Its drought-tolerant properties would have made pearl millet an ideal choice during the arid conditions that prevailed beginning around 1400. Farmers may have cultivated both early- and late-maturing pearl millet varieties, although this awaits archaeobotanical confirmation.[8] Early-maturing varieties are still cultivated in arid northern Ghana today, and seem to have been cultivated along the coast in the early Atlantic era (see above). This means they may well have been present in Banda, though they were probably replaced by maize later on. Whatever the case may be, mixed farming of pearl millet along with the other crops discussed below would have been an appropriate strategy for managing arid conditions and producing a variety of nutritious crops. Mixed farming may also have facilitated experimentation with new crops like maize, which could be easily slotted into field laboratories with little change in agronomic practice.

What did the wider foodscape look like? What crops were being grown, and what did people eat on a daily basis? What did these foods taste and look like? Before we answer these questions, it is important to note that some crops are more visible than others in the archaeological record. Yams and other tubers are almost impossible to find, even though they were probably important staples in the past. Grains and other hard-seeded plants that come into contact with fire are usually the best represented, and this is certainly the case in Banda. But even these grains are not as abundant as at archaeological sites to the north. I suspect this has little to do with the quantities in which they were used, but more to do with the kinds of foods they were made into. In particular, if grains are ground into flour before they are exposed to fire, very few whole grains will survive to be identified under the microscope. The limited written documentation from this period suggests many, if not most, grain-based foods were made from flour on the coast (de Marees [1602] 1987; La Fleur 2012) as well as throughout the drier savanna and Sahel (Lewicki 1974). Lewicki specifies that grains were made into porridges, flatbreads, and fritters (44–49). In Banda, which is situated geographically between the two areas just mentioned, grinding stones are commonly encountered in archaeological deposits from this phase, suggesting that this food preparation technique was quite common.

In order to account for these taphonomic issues, I quantify seed remains in two ways. Ubiquity, measured by the percentage of contexts analyzed where a plant was found, provides an idea of how commonly the plant was used.[9] Percentage frequency, or the percent of the total grain assemblage that a particular grain occupies (e.g., millet in relation to all grains), allows for comparison of one grain type to another, assuming some similarity in preparation. Further, where we found these seeds is almost as important as their ubiquity or frequency, and details are provided in appendix B. I focused analysis on similar kinds of contexts at Ngre and Kuulo Kataa to facilitate comparison. At each I analyzed a midden mound (KK Mound 101; NK Mound 8) dense with garbage and accumulated over a long time period, in order to capture the range of foods used over time; as well as two structures from each village (NK Mound 7 Upper and Lower Structures; KK Mound 118 and Mound 148), in order to begin looking at similarities and differences in peoples' daily lives.

People primarily depended on pearl millet (44% average ubiquity, 85% of total grain assemblage), which has a nutty, earthy taste.[10] While we can surmise, based on the presence of grinding stones, that pearl millet was probably ground into a flour, it is less clear from the evidence how the flour was prepared. Did people enjoy a dish similar to modern day *tuo zafi*, a stiff, polenta-like concoction served with sauce (chapter 5)? While this is a tempting conclusion to draw, we must recall the great spans of time and historical processes between today and the early Atlantic era, a time of considerable flux propelled by innovation and globalization. We know for example that foods like *kenkey*, a fermented maize product wrapped in leaves, were developed on the coast at this time. A close reading of historical sources raises the possibility that people ate pancake-like flatbreads made of pearl millet. Prepared today on hot rocks by mobile pastoralists, flatbreads were more widespread in the past. Referring to the coast in the early seventeenth century, de Marees (1987, 112) characterizes pearl millet specifically as

> a good and excellent Grain, which is turned into bread without difficulty, since it is not hard to break and is quickly ground into Dough. If they knew how to bake it nicely, it would look like and have the colour of rye-and-wheat bread; but as they do not use any Ovens and only bake on the cold [= bare] earth with hot ashes, it looks rather like buckwheat Cakes. It has a good taste and is wholesome to eat. It tastes salty, but grits your teeth a little, which results from the stones with which it is ground.[11]

This description suggests that a millet bread, most likely a flatbread given the absence of gluten, was prepared in southern Ghana in the early seventeenth century. Flatbreads are also mentioned in medieval Arabic sources from Sahelian West Africa (Lewicki 1974, 44–49). Based on pottery forms and the presence of earthen ovens, McIntosh (1995) suggests that flatbreads were the dominant food preparation technique in Jenne-Jeno from 400 BC to 500 AD, but that the subsequent assemblage (500–1500 AD) is more suggestive of whole grains or cracked grains and stews. She interprets this shift as essentially a change in women's labor that necessitated devoting less time to food preparation. While we do not find clear examples of ovens in Banda, note that these were not necessary for bread preparation on the coast. There are several examples of burned basins which could have served as griddles suitable for flatbread preparation (NK M7 Lower Structure and KK M148), with KK M148 perhaps showing evidence for an oven. Phytoliths, microscopic plant silica structures, suggest at least a moderate probability that pearl millet was indeed used in some of these contexts (NK M7 Lower Structure; see Logan 2012 and appendix A for phytolith identification methods).

However it was prepared, pearl millet would have been a nutritious choice, and given its high drought tolerance, also a good agronomic choice. In fact, Banda normally receives too much rainfall to support optimal pearl millet cultivation, so drought during this time may actually have helped increase production. Yet the wider regional distribution of pearl millet remains suggests a more complicated scenario, as seeds are found in even the wettest part of the subcontinent (e.g., Kahlheber, Bostoen, and Neumann 2009; Kahlheber et al. 2014) in places far too wet for optimal pearl millet cultivation. This, along with the grain's long history in West Africa, suggests a strong cultural preference for this crop even in suboptimal growth zones. Such preference is attested to in Banda as well, as pearl millet dominates even the wettest phases that bookend the Columbian Exchange (for the Ngre phase, AD 1230–1400, see Logan 2012; Logan and Stahl 2017; for the Early Makala phases, 1770s–1820s, see chapter 3), when high precipitation levels would have posed serious obstacles to production.

Sorghum, a large globular grain that is sweet to the taste, is also present in the plant remains, but in few contexts and in minute amounts (3% average ubiquity,

4% of the total grain assemblage). While it is present at both Ngre Kataa and Kuulo Kataa, the fact that sorghum's distribution is small suggests it was uncommon across the time range encompassed by these village sites. Sorghum may have been ground or pounded into a starchy staple or perhaps transformed into beer. Beer was probably quite common at Kuulo Kataa, given the preponderance of globular jars with characteristic interior pitting likely caused by the fermentation of an alcoholic beverage (Stahl 2001, 125; Stahl 2018b). Ethnographically, most beer in the area is made of sorghum, but we cannot be certain this was the case in the past. Whatever sorghum was made into, it was probably prepared in individual households, as evidenced by phytoliths likely originating from sorghum in both the upper and lower structures of NK Mound 7.

Maize can have a sweet taste profile like sorghum, though the varieties likely available at its introduction (flint and flour; see Miracle 1965) are much chalkier than the sweet corn commonly found in today's US markets. Maize is present in slightly greater amounts than sorghum (11% of grain assemblage), but with a narrower distribution, as it is only found at Kuulo Kataa, and only in one household (KK Mound 118; 6% average ubiquity overall). Out of the four households tested, KK Mound 118 stands out, since it is also the only one that yields evidence of ivory objects being manufactured on site (Stahl and Stahl 2004, 95), and since its inhabitants seem to have acquired at least part of their food through trade or exchange (see below). One maize cupule from KK M118 was directly dated via accelerator mass spectrometry (AMS) to AD 1484–1660 (at 2 sigma, or 96% confidence level). Documentary evidence suggests that maize did not arrive on the Ghanaian coast until 1554 (Alpern 1992, 2008), so the Banda maize likely dates to the later part of the date range provided by AMS. Still, this demonstrates the relatively rapid movement of the crop approximately four hundred kilometers inland. Maize might also have arrived via an overland route beginning in Senegambia (La Fleur 2012, 95), which makes sense since most of Banda's connections appear to be directed northwards at this time; few material remains are found signaling coastal connections. Combined, these data suggest a quick but limited adoption of maize in Banda, the implications of which I consider below.

In addition to these staple grains, people also consumed vegetable sources of protein and fat. Cowpeas (black-eyed peas) would have provided a complementary protein. Beans do not preserve as readily as grains (Gasser and Adams 1981), so it is not a surprise that they are present in low quantities in Banda. This also explains the presence of seeds conservatively identified as members of Fabaceae, the legume family, which includes poorly preserved cowpeas and possibly another species of bean. Oils would have been important for fat, flavor, and cooking. The oils available in ancient Banda provided rich, robust flavors: true to its common name, shea butter does have a buttery texture, and when fried provides a savory, aromatic quality to food; palm oil is even richer, velvety, and bright orange in color. Shells from both shea tree nuts and oil palm nuts were found, but in frustratingly small quantities of only one fragment each. Given the dry conditions at the time, it is highly unlikely that oil palm could have grown locally (today, under much wetter conditions, it does not); the palm oil or palm nuts were probably obtained via trade. The lack of shea nut shells is more surprising, since these trees do thrive in drier savannas and are often well represented at savanna and Sahelian archaeological sites (Gallagher, Dueppen, and Walsh 2016); their absence points to the collection and production of this oil elsewhere but possibly still somewhere within the Banda region itself. While not cultivated in the strict sense (Gallagher, Dueppen, and Walsh 2016), shea tree stands are encouraged today, and people will travel some distance to collect the nuts.

The role of other wild plants in Banda peoples' diets is difficult to deduce, since most plants that are edible are also medicinal and/or common weeds (see Abbiw 1990). Today for example, leaves from weedy plants are used to make a nutritious soups and sauces that accompanies starchy staples. Several plants used for their edible leaves today (appendix C) were found in archaeological contexts, including *Cassia* spp. (Nafaanra: *bombo*), *Ocimum* sp. (Nafaanra: *napun* [*O. basilicum*] or *chasigbɔɔ* [*O. gratissimum*]), and *Laportea aestivans* (Nafaanra: *klakokagbɛɛ*). Interestingly, *Cassia* spp. are used today to impart a desirable slippery texture to soups, one which allows *tuo zafi* or *fufu* to slide easily down one's throat without chewing. That this texture was desirable is attested to by the presence of okra in both the earlier Ngre and later Makala times (appendix B; Logan and Stahl 2017). However, it is impossible to say with certainty whether these plants were eaten or simply grew nearby.

Sauces or soups may have been flavored by *Ocimum*, a wild basil, as well as by grain of paradise (*Afromamum melegueta*), which provides a peppery, cardamom-like taste. Grain of paradise, or melegueta pepper, was a major trade good at the time, probably originating on the coast and highly desired in Europe from the fifteenth century on (Beichner 1961, 305; Van Harten 1970, 208). Its earliest mentions in European records predate the Atlantic trade (Van Harten 1970, 208–9) and thus suggest that it was initially obtained through trans-Saharan networks rather than coastal ones. This spice is rare in Banda, but the fact that it is present at Ngre Kataa as well as Kuulo Kataa may hint at its role in local cuisine.

Male and female visitors alike might have been offered pipes, since smoking tobacco was rapidly gaining in popularity. Like maize, tobacco was a post-Columbian introduction to Africa, hailing originally from the Americas (McIntosh, Gallagher, and McIntosh 2003). Much like maize, tobacco made fast inroads, and we have good evidence that Banda's inhabitants took quickly to smoking. Smoking is highly visible in the archaeological record, since West Africans were quick to fashion distinctive, locally made ceramic smoking pipes which readily preserve. Documentary, botanical, and archaeological evidence suggest multiple introductions of tobacco in the late sixteenth century. Tobacco is recorded at Whydah, on the Atlantic coast, in 1580, and in Timbuktu in 1594/96, suggesting a much earlier introduction in Senegambia. It was common at coastal Elmina by 1639/45 (Alpern 1992, 30). The variety traditionally grown in West Africa (prior to large-scale commercial production) is *Nicotiana rustica*, which is native to eastern North America; this along with other evidence suggests a French introduction through Senegambia (Ozanne 1969; Phillips 1983).[12]

In Banda, people begin using tobacco pipes in the mid- to late Kuulo phase, with dramatic increases by 1600–1700, indicating a considerably more widespread adoption than maize (Logan and Stahl 2017; Stahl 1999b, 2002). This is particularly impressive since adopting tobacco meant that people also adopted the entirely new practice of smoking; unlike maize, tobacco could not be easily slotted into existing practices. At its first introduction, some individuals at Kuulo Kataa used a variety resembling *Nicotiana rustica*, judging from the morphology of the seeds (in Mound 118, *Nicotiana cf. rustica*: 4 and cf. *N. rustica*: 1; in Mound 148, cf. *N. rustica*: 1). This restricted pattern hints at a quick but limited adoption, like maize, but unlike maize, tobacco seeds are tiny and do not preserve as readily. The distribution of tobacco pipes is likely a more robust indicator of the distribution of smoking.

LOCAL GRAINS AND FOOD SECURITY DURING DROUGHT

This world of interconnected and productive craftspeople and traders attested at Banda does not fit easily with imaginaries of an Africa plagued by the worst drought on record for the last millennium. While this drought surely had an impact on crop production and thus food availability, we do not see evidence of coping mechanisms that indicate food insecurity. Maize, for example, was not adopted widely, suggesting that there was little need for additional calories to bridge a hungry season. While wild plants continue to be used, none of these are known famine foods or replacements for staple crops; instead they added taste and texture. The same is true of the wide range of animal species utilized, some of which seem to have been selected for their novelty or rarity rather than ease of capture, more typical in later periods. Indeed, many of these animals, such as the great white shark or the various predatory cats, were likely luxury items. The production of crafts beyond local needs, including luxury items crafted of ivory and copper, is not what one would expect if drought had resulted in widespread crop failure or food shortage. Neither are the imports of other fine goods like beads and the new habit of smoking, or the use of surplus crops for beer production. Farmers clearly adjusted what was being grown to better suit the dry conditions. Pearl millet, a very drought-tolerant crop, was grown and consumed in higher percentages than sorghum.

However, the question that remains is whether locally produced foods were sufficient to maintain Banda's diverse population. The capability of local food production to meet local food consumption needs can be addressed by considering both whether food was traded into Banda and the local population size. Unfortunately, it is impossible to reconstruct the exact quantity of grains that were produced in Banda, but two lines of evidence help us answer this question. The first is whether grain was grown locally or had to be acquired through trade. We can look at whether crops were being grown locally through tracking crop byproducts. "Byproduct materials" is a general category that includes all non-seed, inedible plant parts from domesticated grains, primarily glume, bract, and rachis material left over from later stages of processing. In grains, these materials vary in their robustness and their ability to be preserved; sorghum and maize have hard glumes and cupules that preserve nicely when charred. Pearl millet has soft, papery glumes that do not stand up as well to the ravages of charring, making specific identification difficult. These preservation differences make it hard to compare between crops (say maize to pearl millet), since the hard-glumed species will always be overrepresented. But they do tell us that plant processing was occurring in those contexts. The reasoning here is simple: grain was likely traded in relatively clean form, since, especially in premodern contexts, it makes little sense to transport extra bulk.

Villagers thus appear to have been growing their own grain or obtaining it from nearby, though this varies between households. A large trash dump from each village was combed for plant remains, and results indicate that pearl millet was very common, occurring in 93 to 100 percent of midden contexts. Chaff, rachis, and other fragments left over from grain processing were present in every midden and house, but in variable amounts across contexts, with the highest ubiquities observed in Kuulo Kataa's midden (40% presence, Mound 101) and a household at Ngre Kataa (32% presence, Mound 7 Upper Structure). This is significant when you consider that most of the grain people were eating was pearl millet, and that pearl millet byproducts preserve the *least* well compared to sorghum and maize. At least some households were processing grains themselves, as suggested by the relatively high ubiquity of pearl millet (32%) byproducts in the Upper Structure of Mound 7 at Ngre Kataa. This is contrast to the household buried in Mound 118, where only 5% of contexts contained byproducts, suggesting that this household acquired much of their grain through trade.

These household-level differences suggest grains were traded between households or within the Banda region. This strategy makes sense of the data at hand. We have plenty of evidence that various craft specialists (potters, metalworkers, ivory workers) lived and worked in the region. Archaeologists tend to make the assumption that the presence of full-time specialists implies a certain level of surplus, since not everyone has to take the time to farm.[13] In Banda, we cannot tell for certain whether people practiced their crafts full- or part-time (perhaps alongside farming, as is common today), but the differences in byproduct remains between households suggests that at least some may have supplemented household food supplies with grain from others. While it is impossible to say with certainty, it is likely that foodstuffs like grains were traded between households and on a regional basis. Ethnohistorically there was a strong tradition of exchanging pots for grain, whereby a pot could be obtained by exchanging a volumetrically equivalent amount of grain. We know from Stahl et al.'s (2008) sourcing study that people obtained pots from different specialists mostly east, but occasionally west, of the Banda hills, and it is quite likely that grains were a part of this regional trade system. In short, it seems likely based on the evidence at hand that the Banda region produced enough grain to feed people, but that grain was traded between households and towns, a social strategy that probably enabled a degree of resilience to local-scale environmental perturbations.

A second line of evidence concerns population: Was production sufficient to support significant populations in the Banda area? This line of reasoning is loosely based on Malthusian assumptions, and is the most common question asked when I deliver talks on food security during the mega-drought. Many people surmise that populations must have been much lower during the mid- to late Kuulo phase in order for diminished food supplies to have been sufficient. Precise population numbers are beyond the reach of most archaeologists, but we can compare the number and size of archaeological sites in one time period with those of another, allowing for a relative assessment of population over time. The Kuulo phase was characterized by aggregation of populations into larger towns (Smith 2008). In fact, there are more large towns on the landscape than in any other archaeological phase (though fewer than today). As later chapters demonstrate, our first evidence for chronic food insecurity is in the mid- to late nineteenth century, when populations appear to be at their lowest (chapter 4). Clearly the predicted relationship between population and food supply does not explain food insecurity in Banda. Instead, the data suggest a far more intriguing possibility: that despite the worst drought on record in a millennium, Banda farmers sustained the largest population in one thousand years.

Banda's case presses the question of whether other ancient African societies were equally well equipped to deal with drought. Unfortunately without equivalent empirical data from archaeological and historical sources, this hypothesis is impossible to adequately assess at present. However, Arabic sources from the interior, drier reaches of West Africa provide tantalizing clues. Lewicki's (1974, 22–24) compilation of Arabic sources mentioning food suggests that food was abundant. For example, Al-Omarī, who wrote between 1342 and 1349 during the height of the Mali Empire, writes that rice, *fonio* (hungry rice, *Digitaria exilis* and *Digitaria iburua*), wheat, and above all sorghum provided ample food for people and animals. He states that this was the case even though the country suffered from several years of drought and continued to supply large quantities of livestock as offerings. Leo Africanus (1526, in Lewicki 1974) found Timbuktu, then a principal city of Gao, as well as Hausaland to the south to be rich in grain. Clearly much work remains to be done to evaluate these claims, as well as to establish the relative food security in these important trade centers. Banda's case suggests that we start with a hypothesis of abundance rather than one of privation, and opens up new possibilities for research into African food security across the wider region.

Banda's situation provokes the question of whether people were able to achieve not just food security but food sovereignty. The term *food sovereignty* in a modern context refers to peoples' rights to foods (as a human right rather than a commodity) and to their control over food resources, often on a regional scale. At present, food sovereignty in Banda during this time is difficult to measure based on the archaeology alone, but the diverse economy of Banda and Africanist scholarship on wealth and value provides some hints. Guyer and Belinga's (1995) wealth-in-people model established that in some areas of precolonial Africa, people, rather than accumulation of goods, were the main source of wealth. People with a variety of skills sets—potters, iron workers, artisans skilled at working ivory, et cetera.—would have been especially valuable because they would have increased the composition of skills. Now, we cannot know how extensive these systems of value were in precolonial Africa or whether they applied to Banda, but they provide a more appropriate model of wealth than do Western capitalist accumulation models. To take the model a step further, food supplies were likely adequate to support the diverse community of craftspeople in Banda. To ensure that people stayed in a region, mechanisms may well have been in place to distribute food even to people who were not primary producers. If not, people of diverse skills sets could have easily voted with their feet (see also Richard 2017). Whether this "moral economy" and the underlying political organization provided food as a right of citizenship is a question worth asking of archaeological data in the future.

RETHINKING THE COLUMBIAN EXCHANGE IN AFRICA

How can we now understand crop adoption during the Columbian Exchange in Banda, given that food security seems to have been high? Much like scholars of the Columbian Exchange, archaeologists seek to understand why people took new plants and animals into their agricultural and food regimes (Jones et al. 2011). This debate is often framed in terms of necessity versus choice: Did people adopt new foods because they were desperate for them, or in a situation of plenty? This question has only rarely been considered in African contexts, where necessity has often been assumed, as Crosby's argument demonstrates.

Macbeth and Lawry (1997, 4) suggest that foodways tend to change the most under the opposing conditions of novelty and of necessity, with less change observed in the vast middle ground. This range of food choice is visible archaeologically through considering the context of food adoptions and their patterns of adoption. Crops adopted out of need to make up for a shortfall tend to be taken up quickly and widely (Liu et al. 2014), whereas foods adopted in conditions of luxury are adopted quickly but in a more limited way, by those with the means to do so (Boivin, Fuller, and Crowther 2012). In the case of the Columbian Exchange in Africa, most scholars have assumed that maize was adopted under conditions of deprivation, since the potential of indigenous African grains has long been undervalued. However, the pattern of maize adoption and spread in Banda does not support this model. If maize was adopted to fill a caloric shortfall, we would have expected a quick and widespread pattern outlined for other instances of food globalization (Liu et al. 2014). What I have shown is precisely the opposite: during the early years of the Columbian Exchange, Banda was in its heyday and maintained a high level of food security. This security enabled a much greater control over food choice, perhaps supporting a degree of food sovereignty. Maize shows a quick but limited pattern that suggests its adoption was a social process (cf. Boivin et al. 2012). Tobacco too seems to conform to this model at its earliest introduction, though it quickly became widespread, suggesting uptake by the masses, out of not necessity but desire.

Understanding context, and particularly food security levels, is essential to disentangling why people adopted new foods, as I discuss further in chapter 3. Mound 118, for example, seems to have been elite, particularly given its early adoption of maize. But objects acquired from long-distance trade are peppered throughout the other houses and trash deposits. Copper and beads, both acquired from Niger trade networks, are found in all structures. Figurative weights that are ethnohistorically associated with gold-dust trading are found in two structures (Ngre Kataa's M7 Upper Structure and Kuulo Kataa's Mound 118). Cowrie shells, used either as currency or for decoration, are nonlocal and are found in three structures (Banda Research Project 1995, 2009). The point here is that the residents of Mound 118 are hardly unique in their desire for global goods, nor do they represent outlier consumers of luxury goods. While it is interesting that they were early adopters of maize, what is more interesting is that most people were not.

If we focus only on change we miss what the majority of people were eating and why. The data from Banda are unequivocal: pearl millet was the staple grain, before, during, and after the Columbian Exchange and until relatively recently. This crop likely enhanced people's ability to withstand climatic risk in a way that maize never could. It is archaeology's ability to access these everyday choices (Robin 2013) that makes it such a critical archive for understanding food history. While archaeologists tend to privilege change, this obscures what actually went on in the past by making those occupying either the luxury or necessity ends of the spectrum the most visible. But people who fall somewhere in between are perhaps making an even more important choice from the vantage of food security: keeping things the same. Continuity is work, most especially in global encounters, when new foods, things, and ideas provide a new set of options. The desire to keep things the same may well arise in response to global encounters, just as local foods movements today are linked to increasing globalization (Wilk 2006b). Such continuities form the basis of tried-and-true traditions that help people weather environmental change and provide a local sense of identity in the face of rapid political economic change. Simply put, if we focused only on change in the form of new crops, we would miss precisely what allowed people to survive the worst drought in the last one thousand years.

3

Tasting Privilege and Privation during Asante Rule and the Atlantic Slave Trade

Waves of turbulence greeted Africans on the eve of the eighteenth century. What would later become first Britain's Gold Coast protectorate then its Gold Coast colony was subject to two major forces that were to have dramatic and sometimes devastating effects: the ramping up of the trans-Atlantic slave trade and the rise of the Asante state. While some areas were depopulated, others benefited from the increased wealth associated with Atlantic trade. Based in Kumasi, the Asante Empire was to rapidly expand into much of modern-day Ghana, creating opportunities for some and resulting in captivity for others. People privileged enough to control or fall outside of these turbulent tides witnessed fewer extremes, but were still affected by the changing fortunes of their neighbors.

Food was central to the experience of these events. Great food stresses for some enabled food excesses for others. From the eighteenth century onwards, pronounced inequalities developed across many West African societies involved in the slave trade, cleaving Africa's economic future on a different path from that of Europe (Green 2019). Yet inequality and its role has largely been overlooked in discussions of African food practices during this period. Instead, the dominant narrative has been a Malthusian one, where the adoption of American crops fueled population increases that led to the rise of complex polities like Asante (McCann 2005; Wilks 1977, 1993, 2004). These purported population increases are even hypothesized to have offset the losses suffered as a result of the Atlantic slave trade (Cherniwchan and Moreno-Cruz 2019; Crosby 2003; Curtin 1969). In this chapter, I critique these arguments and offer an alternative view that focuses on the varied food experiences of West Africans from the late eighteenth through the nineteenth century as a means to highlight inequalities in the food and political landscapes.

This chapter's culinary odyssey begins in the Asante core, where the Asantehene (paramount chief) and his retinue displayed vast quantities of wild and domestic animals, tubers, and grains that boasted of their highly productive agricultural system and control over the countryside. For the remainder of the chapter, I use Banda's experience as a touchstone to explore the range of situations outside Asante's core. At various points in the eighteenth and nineteenth centuries, Banda's inhabitants tasted the full spectrum of privilege and privation through periods of relative stability as well as captivity, all reflected in and experienced through their foodways. In exploring this continuum, we can coax out the situated contexts in which scarcity emerged. The resulting cartography of food experiences compels us to think through diverse African agencies and inequalities as an antidote to homogeneous representations of a scarce Africa.

AMERICAN CROPS AND WILKS'S BIG BANG THESIS

In the early eighteenth century, central Ghana's political landscape was profoundly reshaped by the rise of the Akan-speaking Asante state and its subsequent expansion to much of modern-day Ghana (Arhin 1967a, 1967b; Wilks 1975). To Ivor Wilks, preeminent historian of Asante, food supply was critical to Asante's emergence. He posited that there was a "Big Bang" in the forest zone of Ghana in the fifteenth and sixteenth centuries, triggered by the development of agriculture and the acquisition of what he termed "unfree" labor. This new subsistence regime was fueled by the arrival of American crops and set in motion major transformations in relations of production as well as population increases (Wilks 1977, 1993, 2004). The Big Bang was financed by the development of the gold trade, which allowed forest dwellers the capital to secure unfree labor. Unfree laborers were especially critical in initial clearance of the forest due to the prohibitive labor costs of converting primary forests into agricultural fields (Wilks 1977).

While Wilks's argument has been the subject of numerous critiques (e.g., Chouin 2012; Klein 1994a, 1994b, 1996; Pavanello 2015; Shinnie 1996), the Big Bang thesis remains generally accepted among many Africanist historians (e.g., Chouin 2012, 17; Miller 2015–16) and beyond, even appearing in a recent *New York Times* article on African cuisine.[1] In this section, I focus on the elements of Wilks's argument that specifically relate to African agriculture. The reasoning used to make these claims shows how contradictory evidence is often overlooked even by remarkable scholars, and how those evidentiary gaps are implicitly filled in with tenets of the scarcity slot. As we will see, Wilks's thought process was in part a product of the limits and affordances of the archives he relied upon, as well as of misconceptions that were—and to some extent remain—common among archaeologists and historians. I zero in on the Big Bang precisely because Wilks has clearly laid out his logic in multiple publications, for which he is to be commended. The Big Bang thesis is in many ways a highly conjectural outlier to Wilks's otherwise strong scholarship. This difference reveals the degree to which it is acceptable to speculate about the insufficiency of African agriculture where similar arguments about other skill sets (i.e., weaving, iron-making) would not be deemed credible.

Wilks's (1993, 2004) ideas regarding the origins of agriculture and settled life were rooted in a clever reading of Asante oral traditions and archaeological evidence. His central argument rested on the claim that there was no agriculture (Wilks 1993), or as he later modified it, no fully formed agrarian order (Wilks 2004) in the forest prior to the fifteenth or sixteenth century. He cites Asante origin myths that tell of hunters founding new sites for villages in the forest which were subsequently cleared so that farmlands could be established (Wilks 1977; 1993, 64–66). He tethers this myth to Oliver Davies's archaeological work near Kumasi, which located polished stone chisels that Davies argued were traded with early farming communities. Wilks concedes that this evidence suggests a degree of sedentary living and perhaps some cultigens in a subsistence regime reliant mostly on foraging. He is describing what archaeologists know as the Kintampo Complex, an early farming tradition in Ghana that dates to 3,000–3,500 years before present, long before it is possible to definitively trace the origin of Akan-speaking peoples. While Asante origin stories do accord hunters a central role in finding new villages, it is curious that Wilks interprets this as evidence of the existence of pre-agricultural hunter-gatherers. In Africa, hunters have long played central roles not only in hunting and gathering but also in agricultural societies (e.g., Dueppen and Gokee 2014; Gautier and van Neer 2005; Logan and Stahl 2017; Stahl 1999b; see also chapter 2), a role some still play today. Their range is wider than that of most villagers, since they can be gone for days or weeks in pursuit of animals, so it makes sense that they would act as scouts for new locations. Retracing Wilks's logic reveals the influence of the nineteenth-century writings of Thomas Bowdich, a young British merchant who visited Asanteland in 1817–18 and wrote extensively of his experiences, as well as perhaps the influence of the social evolutionary archaeological reasoning common at the time of Wilks's original exposition in the 1970s (Wilks 2004, 42–52).[2] Curiously, he continued to insist upon this interpretation even after clear archaeological evidence became available indicating the existence of agriculture from 3,500 years ago to the present.

Recall that the period in question is the fifteenth and sixteenth centuries AD, over two millennia *after* we have firm evidence for the development of food production and settled village life in central and northern Ghana (Casey 2000; Chouin 2012; D'Andrea and Casey 2002; Klein 1996; La Fleur 2012; Shinnie 1996, 201; Shinnie 2005; Watson 2010; Vivian 1992), including in Banda (Stahl 1985, 1986, 1999b, 2001). By 2004, Wilks was aware of some of this evidence, but nonetheless argued that agriculture did not cross the forest boundary for two millennia, despite evidence for large towns in Asanteland by 800 AD (Vivian 1992, 161–62) that would have required an agricultural base. His reasoning relates to the high labor costs of converting primary forest into agricultural fields; the unsuitability of savanna crops to forest conditions; the relative unproductivity of yams; and the need for oil palms to be tended (Wilks 2004, 49–50). While I discuss the specific merits of some of these points in a moment, there is a larger issue at stake here: the assumption that Africans were incapable of significant environmental modification until very recently. He employed similar reasoning to argue, for example, that the spread of maize into the forest coincided with a drought that would have facilitated clearance (Wilks 2004, 65–66), rather than crediting sixteenth and seventeenth century farmers with the appropriate labor or skills to make this shift.[3]

Similar reasoning permeates many archaeologists' models of the earlier expansion of Bantu-speaking peoples from west-central Africa to southern Africa, about two thousand years ago. Active debates rage as to whether climatic-induced forest openings or anthropogenic clearance permitted the expansion of these agriculturally based peoples (compare Bostoen et al. 2015 with Garcin et al. 2018). Pearl millet is a savanna crop that is ill-suited to forest conditions, yet its consistent presence across archaeological sites in Central and West Africa suggests that farmers possessed the ability to modify their techniques to forest settings (Garcin et al. 2018). In central Ghana, the archaeological record is clear: people made intensive use of oil palm using arboricultural methods and used domesticated pearl millet and cowpea by 3,500 years ago (D'Andrea, Logan, and Watson 2006, 2007; Logan and D'Andrea 2012).

The idea that forest and savanna agricultural techniques were fundamentally incommensurate is also an exaggeration. Lying between forest and savanna is a thick band of transitional zone referred to as the forest savanna mosaic or the woodland savanna. Banda lies within the wetter part of this zone, and is characterized by high agricultural productivity in modern times, particularly of yams, sorghum, maize, and until recently, pearl millet. Although the quantity of rainfall received in Banda is less than in the forest, both areas share a two-peak equatorial rainfall pattern. The forest fallow system which is described by McCann (2005, 47–48) as unique to the forest relies on a nearly identical set of techniques and agricultural schedule to that employed by Banda farmers. These techniques were clearly highly developed in Banda by the fifteenth century (chapter 2) and probably by the beginning of the first millennium AD (Logan 2012) if not before (Stahl 1985).

Given this evidence, we must ask why the forest remains such a boogeyman to some Africanist archaeologists and historians. In Wilks's case, this idea may relate to early European sources that "misread the African landscape" (Fairhead and Leach 1996). For example, Wilks cites an anonymous Portuguese 1572 report in which the author was trying to deduce the locations of the gold mines in the forest, which local Africans did their best to obfuscate. Wilks (2004, 62) writes that the anonymous chronicler

> knew something of the environment in the mining districts, "The land," he wrote, "does not enjoy the benefit of the sun on account of the great thickness of the trees, into which animals cannot penetrate, let alone the blacks." But obviously there were

miners . . . The implication of the 1572 Report is that at such sites hunter-gatherers were both prospecting and carrying out mining operations. The writer saw that the spread of agriculture was essential to the development of the extractive industry. "Once the hills have become opened, and the land cultivated," he predicted, "thereafter what is produced will be enormously increased . . . "

Here Wilks is critical of the chronicler's remarks about mining, but appears to have accepted his contention that the land was uncultivated. Wilks was clearly aware of the pioneering research of James Fairhead (he is cited in Wilks's acknowledgments), who along with Melissa Leach demonstrated the multiple ways in which colonial officers misread African agricultural practice (chapter 2). Might a Portuguese chronicler with even less information also have misread—or even completely missed—forest-based agriculture? Tropical agricultural systems mimic the forest canopy to best make use of sun, water, and soil; multiple crops including tubers, grains, beans, and vegetables are all interplanted in the same fields. To European observers used to seeing clearly demarcated, monocropped fields, African agricultural plots must have looked chaotic and unrecognizable as proper agriculture or even as agriculture at all. The invisibility of these fields to the 1572 chronicler is evident in his observation of plantains growing in the forest, "where no one could have planted it" (Wilks 2004, 68). According to Dickson (1964, 25), another major source for this period, Dupuis (1824), seems to have misread the forest landscape in a similar fashion.

Wilks's Big Bang hypothesis was infused with new life by McCann (2005, 44), who argues that maize fueled the rise of agriculture in Asante. Rather than providing evidence for early uptake of maize in the forest, which to my knowledge is not recorded in any documentary or archaeological sources, McCann (2001, 258) points to the apparent synchrony of maize's introduction around 1500, the same time that Wilks proposed for the Big Bang. Yet we must be cautious in interpreting this coincidence, which is also when written sources first become available for the Gold Coast. With a degree of circularity, Wilks (2004) marshaled support for the association of maize with the Big Bang by suggesting maize may have spurred the development of agriculture in the forest. In this he appears to rely on Bowdich's (1819 [1873], 181–82) admittedly preliminary interpretation of twelve Asante lineage names: "Regarding these families as primæval institutions, I leave the subject to the conjectures of others, merely submitting that the four patriarchal families, the buffalo, bush cat, the panther, and the dog, appear to record the first race of men living on hunting . . . . The introduction of planting and agriculture seems marked in the age of their immediate descendants, the corn stalk and plantain branches." Key to Wilks's argument are this evolutionary interpretation by which hunting must precede agriculture, as well as his conclusion that the Akan "*abrootoo*" or corn (grain) stalk is equivalent to the more specific "*aburoo*" or maize. Bowdich, a merchant rather than a botanist or linguist, did not distinguish the specific plant to which he refers, using the British term "corn" to refer to grain in general. Bostoen (2007) notes a similar interplay of words for grain crops in many African languages. Words for the indigenous staples pearl millet, sorghum, and finger millet are often used interchangeably with each other and for maize as well. The limitation of this linguistic specificity make linguistically tracing maize difficult in areas that already have strong grain crop traditions, and highlight the need for archaeological data to investigate the roles of various grain crops.

McCann (2001, 2005) suggests that maize was needed to support diets in the forest that were protein-rich but carbohydrate-poor (an assumption Wilks [2004, 49] also makes, contra McCaskie [1995, 29]). McCann reasoned that pearl millet and sorghum could not have been successful in the forest due to their need for sunlight and a long dry season (McCann 2001, 257), though the evidence cited above substantiates the cultivation of pearl millet in the forest millennia ago. Yet maize also requires sunlight and hence forest clearance. Supporters of the maize hypothesis never address why people would have gone through the trouble to clear forest land for maize but not for the indigenous grains sorghum and pearl millet, which had much longer histories and for which there was a much stronger preference in the sixteenth and seventeenth centuries (chapter 2). Forest soils are also more fertile than savanna ones, even if sunlight is more readily available in the savanna (Richards 1985, 49). Although yams are well suited to growing in forest conditions, McCann (2001) argues they were merely prestige foods given their long maturity and high labor requirements. Wilks (2004, 49) also dismissed the potential contribution of yams to forest diets. Both scholars seem to have overlooked Miracle's (1966) comparisons of the productivity of maize versus other staples. In humid, forested areas, yams produce more yield per hectare than maize when considered in terms of calories, while in the park-like forest belts, yams, millet, and sorghum produce higher yields than maize (Miracle 1966, 206). In terms of calories per unit labor, the comparison is more equivocal, with maize performing only slightly better than yams (Miracle 1966, 210). Given the symbolic and ritual importance of yams to Asante, I suspect this slight advantage was insignificant.

Context- and ecozone-specific details like these are understandably sacrificed in the writing of general histories in order to make larger arguments, so the omission of such details in McCann's *Maize and Grace* is not surprising. In a laudable effort to demonstrate the relevance of African food history to global food history, McCann advances arguments that stem from a more general body of theory on maize's adoption in other parts of the world, particularly work by McNeil that credited maize with increasing populations and sociopolitical complexity (McCann 2005, 40–41). Archaeologists also commonly use this line of reasoning, for example to explain the introduction and spread of maize in eastern North America. McCann (46) draws inspiration from Wilks's Big Bang thesis, arguing that maize "spurred an agricultural carbohydrate revolution that allowed forest peoples . . . to feed a dense growing population and fostered an elite political class, royal courts, and a standing army." Variations of this explanation for maize adoption are very common among Africanist archaeologists as well.

There are logical flaws in the idea that maize spurred a carbohydrate revolution and in turn the rise of Asante. Miracle's arguments suggest that maize did not offer many, if any, advantages in terms of yield. Even if it did, McCann's explanation is ultimately rooted in the Malthusian logic that there is a straightforward calculation between calories produced and mouths fed, an equation that is often discredited once more archaeological data become available and when inequality is considered. Even today, people rarely choose the most optimal or efficient source of calories and instead gravitate towards culturally preferred foods.[4] Calories are also a poor measure of nutrition. And all of these issues are complicated by differential access to food supply—something which was likely important in a context of increasing inequality from at least the eighteenth century onwards (Green 2019). Each of these points is of major relevance in today's food security landscape, and thus it is critical we pay them heed when discussing the past. In order to do so, I attempt to evaluate the hypotheses that the forest was carbohydrate poor, and that maize made up for that lack, by evaluating extant data on Asante foodways. I also outline the food practices of Asante elite and commoners, which serve as a point of comparison for the remainder of the chapter, which focuses on food in Banda as well as among those captured as part of the Atlantic slave trade.

Like Wilks and McCann, I am constrained by an archive that largely focuses on nineteenth-century Asante (McCaskie 1995, 26), long after this so-called revolution would have taken place. With a few exceptions (e.g. Vivian 1992), there has been very little archaeology done on early Asante. I attempt to address this limitation in two ways. First, I review available information on Asante foodways, which date to primarily the nineteenth century, reasoning that if maize was as important as Wilks and McCann suggest, it would still have had an important role at this time. I acknowledge the limitation in this kind of argument, however, since people often employ crops like maize under a specific set of circumstances, and return to more preferred foods after that situation is over. Second, I investigate food in Banda, which by the last quarter of the eighteenth century was part of the Asante Empire. This too is an imperfect comparison, since Banda lies within a different ecological zone and the time period is somewhat later. Recall, however, that chapter 2 revealed a fully functional agricultural system capable of maintaining dense populations, even during drought, spanning precisely the centuries that Wilks credits with early, incipient agriculture in the forest. Although in a different ecological zone, the material record demonstrates dense trade relations with the forest and the northern savannas. Nearby Begho likely engaged in these connections with an even greater intensity. Given these relationships, as well as the similarities between forest and forest-savanna agriculture, it is difficult to justify Wilks's argument that there was an impermeable social and economic border between these two ecological zones.

In addition to evaluating the Big Bang thesis, my aim in this chapter is reframe the narrative by contextualizing human experiences of the eighteenth and nineteenth centuries through foodways. I see this as a necessary corrective to a history that has been dominated by Malthusian assumptions based on crop properties rather than by information regarding people's choices and tastes. Moving from the Asante heartland to its Banda outskirts to the Atlantic coast, what becomes painfully evident is the existence of a variegated food landscape that enabled some to enjoy great culinary excesses built on the extreme deprivation of others.

THE ASANTE STATE

The Asante are perhaps one of the best-studied polities in Africa, aided by a density of nineteenth-century visitor accounts, oral histories, colonial records, and the pioneering anthropological work of R.S. Rattray (1923) in the early colonial period. Political order and trade, particularly from the vantage of Asante elites (McCaskie 1995), were the focus of many of these archives and as a consequence, such topics defined historical scholarship throughout much of the 1960s–1980s (e.g., Arhin 1967a, 1967b, 1970, 1990; Wilks 1975, 1993). Asante history has since been broadened to include women, peasants, and social life (e.g., Allman and Tashjian 2000; Arhin 1983; Clark 1994; McCaskie 1995), among other topics common in social history. Yet for all the possibilities in this dense archive, we still know remarkably little about food and agriculture. For the present study, one of the values of elite-centric histories is that they provide us with a lens into the consumptive practice of people with high levels of access to preferred foods, and thus serve as an important point of comparison to the food experiences of the less privileged. In this section, I provide some basic context for discussion of Asante foodways, since economic positionality relates closely to one's food choices. Asante's expansion and reorientation of trade networks fundamentally changed the contours of inequality in the eighteenth century onwards, ultimately sweeping up Banda and many other polities into its imperial grasp.

The Asante state was and is based in Kumasi, a city situated to the north of Lake Bosumtwi in the forest zone. In the early nineteenth century, Kumasi itself was quite large, with population ranging between 20,000 and 25,000 inhabitants, many of whom were engaged in government business (McCaskie 1995, 33). Many government officials also owned farms that provided sustenance to the urban center and provided a major source of wealth. Those who labored on these farms were generally slaves, wives, children, and pawns (people who were given in service for a specific period as payment for a debt) (Arhin 1990, 527). Surrounding populations brought meat, agricultural products, and *pito* (local beer) to Kumasi markets to supply the urban populations. According to Bowdich ([1819] 1873, 320), market goods in Kumasi included local produce such as staples (yam, plantains, grains, sugar cane, and rice), vegetables (okra, peppers, and garden eggs), fruits (oranges, pawpaws, pineapples, bananas), meat (beef, mutton, wild hog, venison, monkey, smoked snails, and dried fish), and drinks (*pito* and palm wine). Imports from the north and Europe included household goods (mirrors, brassware, sandals, cloth, blankets, pillows, pipes, drinks, salt, tobacco, and calabashes), farm tools like hoe blades, crafting goods (thread and brass), and weaponry materials (iron, lead, flints, etc.).

Foods, crafts, and trade goods generated profits at thriving markets. Since Kumasi was within the tsetse fly belt, cattle and livestock from the north were important and expensive goods, with cattle costing up to six times more in Kumasi than in the northern trade center of Yendi (Bowdich [1819] 1873, 272). Beef, mutton, chicken, and horses were thus luxury goods that only the king and other high officials could regularly afford. Commoners purchased items like salt, cloth, iron, and craft products in markets. Specialist villages close to Kumasi produced craft goods like textiles, pottery, wood carvings, and objects in gold and other metals. Skills that were needed by a wider array of people, such as those of blacksmiths and potters, were distributed more widely throughout Asante. Yet others were closely monitored, particularly the production of state regalia (Arhin 1990, 527–28).

The primary medium of exchange for Asante elites was gold dust, rather than the cowries or metal disks that had been used in earlier periods, but these currencies continued to be used concurrently in the north (Arhin 1983, 473; Arhin 1990; Garrard 1980). Gold was central to the Asante state economy as both an export and for consumption in internal markets. From the sixteenth century onwards, gold trade out of Akan lands totaled over 3.6 million troy ounces (around 112,000 kilograms) per century, not including the amount consumed within Ghana. Gold was also used for the production of paraphernalia for chiefs. Furthermore, the Asante maintained a substantial amount of gold in state coffers—as much as 800,000 troy ounces (around 25,000 kilograms) by the nineteenth century (Garrard 1980, 163–65). In the eighteenth century, the Asante turned from the export of gold to that of slaves, who were obtained through their expansionary military tactics and tribute demands (see below). After the 1807 British abolition of the slave trade, the gold trade resumed, as gold and ivory exports increased to Europe, and gold trade to the north continued (Garrard 1980, 157–58).

The Asante state was involved in trade on several scales (Arhin 1990; Boaten 1970). Interstate or foreign trade was the most important and lucrative. Asante traders supplied the northern polities of Dagomba, Gonja, and Gyaman (map 3) with forest-produced kola nuts and European goods, and in exchange obtained slaves, livestock, salt, iron bars, shea butter, and coarse cloths (Arhin 1990, 528; Boaten 1970, 35). Trade with Europeans required gold, ivory, slaves, and, later on, rubber, in exchange for firearms, lead bars, gunpowder, cloths, alcohol, and salt (Arhin 1990, 528). From the early nineteenth century, the Asante established a virtual monopoly on northern-focused trade through strict control over the distribution of firearms as well as the establishment of customs houses or checkpoints along trade routes, which charged taxes and tolls (Boaten 1970, 37, 40). Commoners were involved in the kola and later the rubber trade, but gold, ivory, and slaves were domain of the Asante state itself. Arhin (1990, 528–29) suggests that more wealth was generated from the northern rather than Atlantic trade since more people were engaged in it and it was based on kola nut. Kola nuts were an ideal trade commodity, since they grew wild near Kumasi and were available to all who were physically able to collect them, yet they were in great demand by Muslims to the north and east (Handloff 1987).

Even though northern-focused trade may have been more important to the average Asante citizen (Arhin 1987, 1990), the trade in gold and human captives to Europeans on the coast was critical to state finance as well as state expansion, particularly as a means to acquire firearms. The Asante court maintained trade relations with multiple European nations who had set up shop on the coast, which sometimes resulted in rivalries among these foreign powers. As hinted above, most of these interactions were marked with the two-way exchange of food and drink (McLeod 1987, 186). The importance of the northern and coastal trade is commemorated in much paraphernalia for chiefs, including foreign metal vessels, umbrellas, weight systems, staffs, and firearms, among other things (McLeod 1987, 184).

Asante expansion began before 1700 and Asante state influence reached its maximum extent by the early nineteenth century, when it included most of modern-day Ghana and southeastern Côte d'Ivoire (map 4; Arhin 1967a; Wilks 1975). Expansion campaigns in the eighteenth century were directed primarily northward, whereas in the early nineteenth century the focus was southwards. In eighteenth-century Asante, succession disputes and expansionary military campaigns dominated the royal agenda (Arhin 1967a). Asante expansion northwards began with the conquest of Wenchi in 1711–12 and Bono/Takyiman in 1722–23 (map 3). Begho may have also been invaded at this time, or might have gradually declined in prosperity and succumbed to internal struggles (Wilks 2005, 18) or dispersed to new markets (Stahl 2001, 149–50). Asante armies turned northwards and invaded Gonja and Dagomba in 1744 (Arhin 1967a, 74).

The spatial organization of Greater Asante included the metropolitan area of the political capital, the inner provinces, and the outer provinces (map 4). The boundaries of each were marked by control posts along major routes (one such post was in Banda, marking the boundary between inner and outer Asante: Wilks 1975, 54). Inner provinces were more tightly integrated with Asante, while control over outer provinces was more akin to indirect rule (Wilks 1975, 62). Inner provinces were considered subject provinces, which had "Asante law and Asante rights" (Christaller in Wilks 1975, 63), and most were Akan or closely related culturally (Arhin 1967b, 76). Subject provinces may have been forced to pay the same taxes as those in the metropolitan region (i.e., death taxes, war taxes, and household taxes), often in gold dust. Inner provinces were at a disadvantage, however, in that they could not raise revenue by raiding areas outside Greater Asante. Instead, they raided other polities in the inner provinces when additional revenues were needed (Wilks 1975, 69–70). Tributaries or outer provinces such as Gonja and Dagomba (map 3) maintained an even looser relationship with the

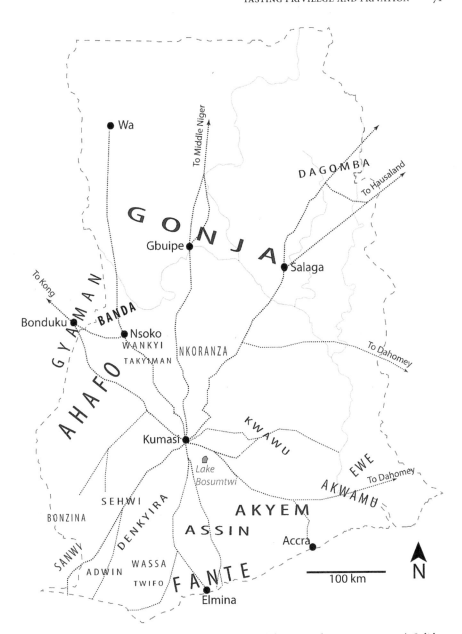

MAP 3. Nineteenth-century polities and Asante roads (after McCaskie 1995, 32, 76, 148). Solid lines are rivers, small dotted lines are roads, and larger dotted lines denote modern-day borders of Ghana.

Asante state. These communities were expected to make a financial contribution in goods or manpower, but their ties with the state were mostly commercial and often of mutual benefit. Asante demanded regular payments from tributaries, but

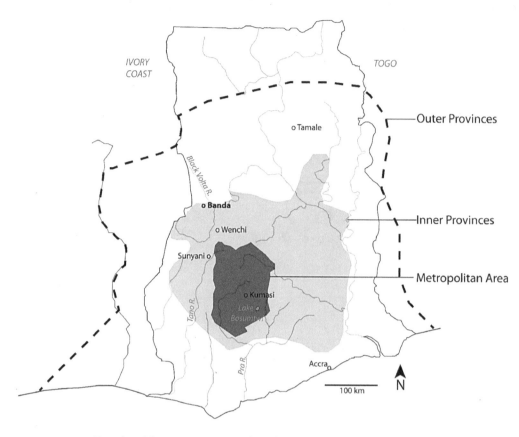

MAP 4. Geopolitical divisions in Asante in the early nineteenth century (after Wilks 1975, 62).

tributaries maintained political independence from Asante. Outer provinces were not required to participate in military ventures with the Asante (Arhin 1967b, 76–77).

The Asante government demanded two kinds of tribute after its conquest of an area. The first was a payment made soon after defeat that was intended to defray the costs of the Asante expedition. The second was an annual tribute requirement. Payment to the state was in money (gold dust, cowries, etc.) or in kind (in slaves or products of the new territory). Payment in products was the norm in areas that relied on cowrie-based rather than gold-based currencies, which included most of northern Ghana (Arhin 1967a, 286–87). Dagomba, for example, paid "500 slaves, 200 cows, 400 sheep, 400 cotton cloths, and 200 cotton and silk cloths," and Takyiman a smaller proportion of the same kind (Bowdich 1873, 269). Yet other communities, such as the Europeans on the coast, paid by means of "notes" or a ground rent of sorts that focused more on providing subsistence, that is, food monies (Arhin 1967a, 287). Tribute demands differed based on the size of the polity and the degree to which it was able to supply gold or other goods. These tributary relationships effectively transferred considerable surpluses from local polities to the centralized Asante state, concentrating wealth and power in Kumasi.

TASTING AFFLUENCE IN THE ASANTE CORE

Food was one of the central ways in which power and status was experienced and displayed in nineteenth-century Asanteland. European visitors frequently detailed the ingredients and presentations of elaborate feasts that no doubt communicated the wealth of the Asantehene and his retinue (Miller 2015–16). Despite these accounts, remarkably little has been written on the food history of the Asante. This is a larger project beyond the scope of this volume. In this section, I detail only a small portion of the available archive as is relevant to the Wilks/McCann hypothesis as well as to the understanding of elite consumption. First, I examine the role of maize relative to other introduced and indigenous carbohydrate sources. Second, I evaluate the relative abundance of food in the Asante heartland by exploring the diets of high- and low-status individuals. This examination of food status displays reveals that maize seems to have played a minor role in Asante cuisine, and that meat was highly valued.

The idea that maize was a mainstay of early Asante diet appears to have originated from two sources: its apparent role in Asante symbolism and its widespread adoption along the coast. The interpretation of maize's role in Asante origin myths has already been questioned in the previous section. Both Wilks and McCann cite an Akan proverb, *"Aduane panin ne aburoo,"* which translates "the chief/elder among foods is maize" (Christaller and Lange 2000). While the proverb is intriguing, no context or etymology is provided and it may well derive from the twentieth century, when maize did become important for a number of reasons (chapter 5). Without additional chronological details, historical linguists consider proverbs as notoriously difficult to date with any confidence (Stephens 2018b, 791), so their deployment as such in Wilks's account needs to be taken with a grain of salt. Historical documents from the coast are more numerous and attest to the rapid spread of maize around European forts (e.g., de Marees 1987). Carney and Rosomoff (2009) have interpreted the quick uptake of maize to its central role in provisioning European forts and ships, an observation I return to at the end of this chapter. What is less clear is the degree to which Africans themselves relied on the crop, particularly in the interior occupied by Asante and other groups (see La Fleur 2012, 150–53 for an intriguing suggestion regarding pellagra).[5]

Other scholars argue that maize played a more limited role in Asante. Both Dickson (1964, 27) and McCaskie (1995, 27) associate Asante's use of maize with its military campaigns, based on the observations of G. A. Robertson in 1818 that the cultivation of maize had been "lately more extensively introduced into the interior of Ashantee; both from its importance in feeding their stock, and being portable for the supply of their armies" (Robertson 1819, 201). To Dickson (1964, 27) the advantages of maize, a portable, storable surplus, were manifold over the main Asante staples of yam and plantain, which are cumbersome and subject to rapid spoilage. The indigenous grains sorghum and pearl millet, however, were not only more widespread in most conquered areas (Dickson 1964, 27), but have a considerably longer shelf life than maize. Bowdich's (1819 [1873], 250) description of soldiers' fare, for example, leaves considerable room for interpretation: "The army is prohibited during the active parts of a campaign from all food but meal, which each man carries in a small bag at his side and mixes in his hands with the first water he comes to; this, they allege, is to prevent cooking fires from betraying their position or anticipating a surprise. In the intervals (for this meal is seldom eaten more than once a day) they chew the boossee or gooroo [kola] nut. This meal is very nourishing and soon satisfies; we tried it on our march down."

Note that the crop used to make the meal is not specified; perhaps this is purposeful, since most grain crops can be ground and eaten in this fashion, providing troops with a considerable degree of flexibility. The preparation method described for soldiers—a ground flour mixed with a bit of water—was until recently a common preparation in Banda (Nafaanrá: *sisa*) using pearl millet or maize and served to men upon their return from the fields. This technique may have been learned during Banda's service in Asante's army (below), and also raises the possibility that other crops were eaten in addition to maize. Even cassava can be made into a coarse meal, known today as *gari*. Asante's soldiers were probably dependent on the areas they invaded for at least part of their provisions, and would have presumably consumed whatever grain was commonly grown in these locations. Asante conquered areas from the dense tropical forest north to the drier savanna woodland, environments which would have posed different constraints on the crops being cultivated. The degree to which Asante soldiers relied on maize likely varied over time as well. Dickson cautions that the Asante may have learned of virtues of maize only in 1806, with the beginning of campaigns to southern Ghana, where maize was grown more extensively.[6] McCaskie (1995, 27) suggests that the popularity of maize waxed and waned with Asante's military campaigns, decreasing between 1800 and 1900. What seems likely from this limited historical data is that maize was a strategic resource used to fill certain aims of Asante, rather than an important source of carbohydrates for the populace.

Even better suited to cultivation in Asante's immediate forest hinterlands was plantain, which La Fleur (2012) argues was the more revolutionary carbohydrate. Plantain was introduced possibly much earlier via the Indian Ocean and eastern Africa. La Fleur (2012, 108) argues that the increased focus on gold, with its attendant labor requirements, may have created a need for a high-yielding crop like plantain, particularly in the forested environments surrounding gold fields. Unfortunately, crops like plantains do not preserve well in the archaeological record, making it difficult to evaluate this thesis; analysis of phytoliths from archaeological sites could be marshaled in the future. However we must be attentive to these archival blind spots, because they significantly constrain our ability to re-create past foodways. Other root and tuber crops like yams and cassava are similarly obscured in the archaeological record, forcing us to rely on documentary and linguistic evidence.

Documentary evidence suggests that cassava arrived on the Gold Coast in the late seventeenth or early eighteenth century, though it was some time before Africans began planting it (La Fleur 2012, 156–57). Cassava was probably introduced into Asante Brong-Ahafo by 1806. Cassava has many advantages over yams, with which it was probably in competition in both fields and cooking pots. Cassava produces high yields in depauperate soils, requires little cultivation, and the tubers can be left in ground for a long time (Dickson 1964, 26–27, 1969, 119). Yams require prime land, are harvested only once a year, and require considerable investments in agricultural labor. Despite these differences in efficiency, La Fleur (2012) illustrates a strong preference for yams and marked dislike of cassava's taste. Cassava was considered poor man's food at least from the eighteenth to the early twentieth century along the coast, and may have been used to make up periodic shortfalls in more desirable crops (La Fleur 2012; McCaskie 1995, 27; see also Ohadike 1981 and chapter 4). In other words, cassava was used strategically when yams were not an option.

Besides the introduction of cassava, Dickson sees little change in what was actually cultivated between 1700 and 1800 on the coast, which according to both Bosman (1705) and early nineteenth-century writers included guinea corn (sorghum), pearl millet, maize, rice, and yams (Dickson 1964, 26). A smattering of other crops were introduced in the eighteenth and nineteenth centuries, including members of the onion family, breadfruit, cocoyams, tomatoes, and avocados (Alpern 1992, 2008). The New World cocoyam (*Xanthosoma mafaffa*), a more palatable cocoyam, had been introduced by the 1840s, likely replacing the other cocoyam, *Colocasia esculenta*, which contains significant calcium oxalate, a digestive irritant (McCaskie 1995, 27). Around 1700, maize had established a niche on the coast but was still secondary to sorghum and pearl millet; by 1800 maize was perhaps more common, though sorghum still dominated in many areas and pearl millet was still grown. Rice was also produced in large quantities on the coast, along with yams, oil palms, coconuts, pineapples, sweet potatoes, bananas, sugar cane, et cetera (Dickson 1964, 26).

Nineteenth-century sources reveal that maize as well as cassava were not a significant part of Asante diet. This point is underscored in Bowdich's ([1819] 1873) observations of diet in Kumasi. He discriminates between white soups and black soups, the latter made with palm oil, though we learn little else of their ingredients. The starch commonly was yam or plantain, at times made into *fufu* (a firm-textured pounded food; chapter 5). Bowdich (267) notes that "they do not make *cankey* [*kenkey*] of their corn (a coarser sort of kouskous not cleared from the husk) as the Fantees do, but they roast it on the stalk, and when young the flavour closely resembles that of green peas." This statement implies that the Asante "corn" (i.e., grain) may have been pearl millet, which was eaten as a couscous, rather than maize, which was commonly consumed along the coast by their southern Fante neighbors as *kenkey*.[7] It also suggests that maize was prepared by roasting on the stalk, rather than as a starchy staple. La Fleur (2012, 94) observes that the forest would have presented challenges to maize storage, since it was particularly susceptible to fungi and spoilage. These constraints meant that people probably favored floury types of maize that could be harvested fresh rather than dried and roasted as a snack or prepared as a vegetable accompaniment to main meal. In sum, the evidence from Asanteland itself suggests that maize was a minor food, painting a much different picture than what has been argued in the Wilks and McCann hypothesis.

If Asantes were not reliant on maize, what did they eat, and how did food vary across social class? Our most detailed archive derives from the accounts of European visitors to the Asante court in the nineteenth century. Quantity and quality of food and drink seem to have played central roles in Asante feasts and in Asante state encounters with European delegations. Upon his arrival, Thomas Bowdich ([1819] 1873, 129) and his retinue were greeted by the Asantehene and conducted to a breakfast reception, "where a relish was served (sufficient for an army) of soups, stews, plantains, yams, rice &c. (all excellently cooked), wine, spirits, oranges, and every fruit." Later that day, dinner was also provided, which apparently exceeded all expectations. Elements of the meal seem to have been prepared specifically for the European audience, including a heightened table as well as silver cutlery. A roasted pig was the central dish, as were

> roasted ducks, fowls, stews, peas-pudding, &c. &c. On the ground on one side of the table were various soups, and every sort of vegetable and elevated parallel with the other side, were oranges, pines, and other fruits; sugar-candy, Port and Madeira wine, spirits and Dutch cordials, with glasses. Before we sat down the King met us, and said, that as we had come out to see him, we must receive the following present from his hands: two ounces four ackies of gold, one sheep and one large hog to the officers, ten ackies to the linguists, and five ackies to our servants. We never saw a dinner more handsomely served, and never ate a better. (130)

Bowdich's experience was one among many, as British delegations in particular sought to establish commercial ties with Asante in the wake of the 1807 abolition of the slave trade. Not all delegations were met with such pomp, as the Asantehene also communicated power through withholding offers of food or drink during hours-long receptions (Sheals 2011 in Miller 2015–16, 37).

Beyond state dinners, feasts also communicated abundance and power through consumption, as is revealed in Bowdich's (226–30) detailed account of the Yam Festival. While he omits what people actually ate, the dynamics he describes say much about hierarchy and consumption in Kumasi and beyond. Social norms appear to have been suspended for all in attendance; as Bowdich (226) reveals that normal punishments were suspended for infractions committed during the festival, where "the grossest liberty prevails." The Asantehene offered copious amounts of rum (an expensive, imported drink) in large brass pans (also imported) for the crowd to drink. Despite spilling more than they drank—another show of excess—people quickly became very drunk. At the same time the brutality and life-and-death power of the state was emphasized by constant musket fire and the display of the severed heads of chiefs who had revolted, as well as the sacrifice of a large number of slaves. Each of the chiefs present sacrificed several slaves, according to Bowdich (228), so "that their blood may flow into the hole from whence the new yam is taken." At the palace, the king also sacrificed twenty or so sheep, pouring their blood as well as palm oil and eggs on family stools and doorposts. After the ceremony was completed, the royal household ate new yam for the first time in the market, signaling that it was time for the populace to enjoy the tuber as well.

We must acknowledge that sources like Bowdich were written by occasional visitors who were shown only a very limited slice of Asante life, primarily that of the nobility and soldiers. We also do not know what festivals looked like when performed outside of the gaze of European visitors, and to what degree the great excesses observed by Bowdich were a performance meant to impress Asante's might upon him. McCaskie (1995, 34, based on interviews) offers us a glimpse of the lavish daily meals of the Asantehene, the paramount chief or king of Asante:

> The Asantehene's household alone daily consumed large quantities of food . . . to-gether with a leavening of imported delicacies. The Asantehene Kwaku Dua Panin (1834–1867) took breakfast about 8 a.m.; meat, plantain and yam "in large quanti-ties" were distributed to the Asantehene's wives and children. The main meal of the day was taken at 2 p.m., when the Asantehene and his household officials made a selection from "mutton, turkeys, ducks, fowls, wild game of all kinds, except the buf-falo; and fish from the lakes, and adjacent rivers . . . also yams, plantains, beans, rice, European biscuits, tea, sugar, wines, liqueurs, etc.". Immediately afterwards, "large dishes containing the great family dinner" were distributed, like breakfast, to the royal wives and children.

The main food of other elites was a "soup of dried fish, fowls, beef or mutton (according to the fetish), and ground nuts [peanuts] stewed in blood," while the poorer classes apparently made do with "soups of dried deer, monkey's flesh, and frequently of the pelts of skins" (Bowdich [1819] 1873, 267). Note here the emphasis in both accounts on animal/meat products rather than starch. Among the starches, in McCaskie's account, yams, which are of well-known ritual and subsistence importance, are mentioned first, followed by plantains, beans, and rice, but curi-ously there is no mention of maize, sorghum, or pearl millet. It is curious that both McCaskie's later informant and Bowdich (in two instances) highlight animal products before starchy ones, and identify these animals in much more detail than the botanical constituents of meals. It is unclear whether this listing reveals the importance of animal versus plant foods in such displays or is an artifact of Euro-pean chroniclers' preferences. This practice might suggest that animal products were of higher value than starches to the court. A similar valuation appears to have been the case along the coast (La Fleur 2012, 134), in Dahomey (Monroe and Janzen 2014), and is suggested by the animals offered as gifts to Bowdich's entourage.

Food archaeologists have long argued that valued foods are often the ones that are the most rare or difficult to acquire, with common foods often assuming a lower status (Hastorf 2003). In Asanteland, available source material suggests that domestic livestock was in shorter supply than carbohydrates. Since much of the empire was below the tsetse fly belt, there was an increased risk of sleeping sickness, making livestock rearing riskier there than in drier locations like Banda and its northern neighbors. Consequently livestock and particularly cattle were highly valued for their meat and hides and were always in short supply in Kumasi (Arhin 1987, 57; Bowdich [1819] 1873, 272). This short supply may have impacted the ability of commoners to access domestic meat on a regular basis. However, there seems to have been ample wild game available in central Asante, ranging from elephants to river fish, and hunting was an established part of the Asante economy. Snails were a particularly important, if seasonal, source of protein (McCaskie 1995, 27–29).

The abundance of agricultural produce expressed in these accounts of the Asantehene's daily meals and his gifts to foreigners boasted of a flourishing agricultural system under state control (McCaskie 1995, 33–35). European observers of nineteenth-century Asante commented on the prevalence and intensity of its agriculture and the abundance of crops produced (Dupuis [1824] 1966; Hutton 1821, 203; Wilks 1993, 49). This was especially true around urban centers like Kumasi, while areas farther afield appear to have enjoyed longer fallows and less intensive agriculture (Wilks 1993, 50). Farther north, the density of farmed land increased as one approached major towns and satellite towns charged with producing crops for the urban areas (Dickson 1964, 27). The highly organized and intensive farming system surrounding Kumasi functioned essentially to supply the state apparatus; their surplus appears to have been large enough to support the Asantehene and his retinue, government officials and members of the elite class, as well as local markets (McCaskie 1995, 33–35). There were, however, real limits on transport of perishable foodstuffs, necessitating state-controlled agricultural intensification by the late nineteenth century through shortened fallows and concentrated labor inputs (McCaskie 1995, 36–37).

It is more difficult to evaluate food security in Asante, particularly among the non-elites, since very little of our historical evidence speaks directly to their food habits. McCaskie (1995, 29) reasons that "deficiency diseases recorded in the 19th century tended to be related to protein shortages, or to failure to maintain a balanced diet, rather than absolute caloric sufficiency, malnourishment or starvation." He bases this finding on notes from a visiting British doctor, H. Tedlie, whose observations and treatments were recorded in a materia medica posthumously published in Bowdich ([1819] 1873, 282–92). While the doctor does mention that the poor and slaves suffered from parasitic and other cleanliness-related diseases, to the best I could tell none of the diseases he recorded are caused primarily by food, including protein, deficiencies. In fact, no European sources attest to severe food shortages in central Ghana until agricultural systems collapsed under the civil war that plagued Asante in the 1880s (McCaskie 1995, 29). Land also appears to have been plentiful compared to population (McCaskie 1995, 30), suggesting high availability of cultivated food products.

In sum, these historical records suggest that Asante's core enjoyed a relatively high degree of food security during the nineteenth century, supported by an extensive agricultural system, as well as hunted and traded animal protein. European observers were impressed by what they saw, and indeed, the resulting historical narratives are excellent sources of information on elite food practices. Yet they leave out people on the margins of society, like pawns and slaves, as well as people occupying the edges of the vast empire. In the next section, I consider how Asante rule impacted everyday life in Banda from the 1770s to the 1820s, using archaeological remains, which speak to food practices at a much different scale than historical records. In the final section, I turn specifically to those who suffered the most: captives and pawns. Cumulatively these food experiences speak to an uneven culinary topography that overturns and complicates a simplified Malthusian view of African foodways.

TASTING IMPERIAL CONTROL IN BANDA, C. 1774–1820S

In chapter 2, I paint a very different picture of agricultural capabilities in the fifteenth and sixteenth centuries than do Wilks and McCann. Banda had a fully formed agrarian order that was resilient enough to sustain even the most intense, prolonged drought on record in the last millennium. And its people did so by relying on pearl millet, a crop which has sustained African populations for millennia. While Banda lies in the wooded savanna some distance north of the forested heartland of Asante, the agricultural regimes of the two regions had much in common. Both areas relied on some of the same crops, though Banda's drier position meant that oil palms and plantains were not widely grown. Both areas also relied on slash-and-burn and multicropping techniques. And there is considerable evidence for communication and active trading between the two zones. For example, several animal species found only in moist tropical forest are recovered in Banda's archaeological record (Logan and Stahl 2017). For generations, Banda was involved in lucrative trade between the forest and savanna that was brokered by nearby Begho, a well-known entrepôt that peaked between the fifteenth and eighteenth centuries (Posnansky 1987).

Begho's decline may well have been associated with Asante's rise, and particularly the sacking of neighboring polities like Bono Manso in the eighteenth century. In Banda, there seems to be a gap in the archaeological record between circa 1650 and 1750 AD, but it is not clear whether this is a real pattern or an artifact of gaps in dating methods (Stahl 2001, 161). Populations may have dispersed during a period of instability in the wake of Begho's decline (Stahl 2001, 161), perhaps as a result of a declining resource base (Goucher 1981) or simply a lack of economic opportunities, as Asante wrested control of trade (Posnansky 1987). Whatever the case, oral histories record the arrival in Banda of a new group, the Nafana, from northeastern Côte d'Ivoire sometime after the fall of Begho and before the arrival of the Asante in the eighteenth century. While there seems to have been plenty of space for this new population, the landscape was hardly empty. The Nafanas encountered the ancestors of the modern-day Kuulo (Dumpo) among other ethnic groups such as the Ligby. In ensuing negotiations the Nafana wrested political control over the area, and the Kuulo retained power over the land via the office of earth priest (Stahl 1991).

The Banda polity emerged before 1751, probably by about 1720 (Stahl 2001, 150–51). In the first half of the eighteenth century, the Asante took control of most of Banda's eastern and southern neighbors. Asante invaded Banda in 1733 in retaliation for the killing of an Asante trader (Arhin 1987, 53), but it was not until attacks in the dry season of 1773–74 that the Asante established formal hegemony over Banda. Oral histories record the traumatic nature of this war (Ameyaw 1965 in Stahl 2001, 155–58). Upon their acquiescence, Banda's chiefs agreed to provide a yearly tribute in sheep. To help offset the costs of war, many young men from Banda were taken as captives to Kumasi, where they were trained in service to "stools," that is, offices (Wilks 1975, 246). Other captives from Banda were sent to the coast to be consumed in the Atlantic slave trade (Yarak 1979 in Stahl 2001, 155; Stahl 2015). In the section to follow, I consider how food security was impacted by these violent encounters with Asante, as well as the experiences of those who were sold as slaves along the coast. In this section I focus on the archaeological record from Banda itself, which represents those who lived under Asante rule in the late eighteenth and early nineteenth centuries.

By the early decades of the nineteenth century, Banda was counted among Asante's inner provinces, and joined the Asante in wars with neighboring polities such as Gyaman (map 3; Stahl 1991, 260; Stahl 2001, 156–58). Asante influence on Banda's political and economic institutions appears to have been quite pronounced, given the similarities in the structure of Banda's chieftaincy as well as the transformations in regional and long-distance trade and craft production (Stahl 1991, 2001), so we might hypothesize that foodways were also impacted. Stahl (2007, 70–71) suggests that incorporation into the Asante empire probably afforded Banda a degree of security and stability. However, this security came at a cost: Banda soldiers were required for Asante's military campaigns, and not only did Asante controlled access to many imported goods, but also, as an inner province, Banda was required to pay taxes (Stahl 2007, 70–71).

The costs of being under Asante control are visible in Banda's shifting political economic base, especially compared to the earlier time periods covered in chapter 2.

Namely, people were not as invested in producing large quantities and varieties of crafts as they had in previous centuries (Stahl 2001; chapter 2). During the period of Asante dominance, women were particularly important in crafting pottery and making thread, while men were known as weavers (Stahl and Cruz 1998). Instead of focusing on long-distance Niger trade, people invested in trade on an immediate and regional scale (Stahl and Cruz 1998; Stahl et al. 2008). This focus on more localized trade networks might have helped build and cement social and economic relationships with neighboring regions. The end of significant long-distance trade may also have been imposed by the Asante, who strove to reroute lucrative trade networks through Kumasi (Arhin 1970; Wilks 1962).

People also constructed their daily lives somewhat differently than they had in previous centuries. While villages were numerous across the Banda landscape, they were somewhat smaller and less aggregated than in previous centuries (Smith 2008, 548). Some chose to live in new villages, and others returned to population centers of earlier centuries like Ngre Kataa. Our best resolution comes from a new village, Makala (Kataa), which was the first settlement of the Nafana people upon their arrival east of the Banda hills beginning in the 1720s (Stahl 2001, 162).[8] At about eighteen hectares or thirty-three American football fields (Stahl 2001, 165) in size, Makala Kataa was a large village, but it was still much smaller than Kuulo Kataa (fifty-two football fields). I analyzed plant remains from two houses buried in Mound 5 and Mound 6 in Makala Kataa. Mound 5 was an exceptionally well-preserved outdoor kitchen, and will be discussed in detail below. Mound 6 included several rooms with far fewer plant remains recovered, attesting to the well-swept nature of interior spaces. A smaller subset of samples was analyzed from two outlier villages, B-112 and A-212 Mound 1, to help provide regional coverage (see introduction, map 2). Full results are reported in appendix B.

Unlike in the preceding phases, environmental conditions were optimal during the time people lived at Makala Kataa. Paleoenvironmental records from Lake Bosumtwi suggest that while periodic droughts occurred, average rainfall was significantly higher than in previous centuries, on par with or greater than in the present day (Shanahan et al. 2009; contra Wilks 2004). Farmers increased the amount of sorghum they cultivated to account for elevated rainfall (28% ubiquity), although pearl millet was more common (45% ubiquity), suggesting continued cultural preference for this grain. Although conditions were perfect for maize production, it remained uncommon in the kitchen (12% ubiquity), despite the widespread use of maize cobs as a decorative roulette on pottery. That people continued to opt for indigenous staples instead of maize despite ideal rainfall complicates the McCann/Wilks Big Bang thesis. Very rarely do people make choices based on optimal yield of calories; rather they prefer to eat those foods that taste the best to them and are part of familiar local cuisines. Because of this long period of experimentation, it seems likely that farmers were well aware of maize's ability to produce two crops per year, yet the need for extra calories or a stopgap crop was missing, suggesting a relatively comfortable level of food security.

Yet the archaeological record also suggests that a range of circumstances and tastes complicated this scenario. People living at A-212 and Makala Kataa Mound 6 did not care for maize at all, to judge from its absence in those contexts. By the same logic, people who inhabited B-112 appear to have disliked sorghum, since it was absent entirely. B-112 is far to the southeast of Makala Kataa, and may have been occupied by different social groups. Oral traditions record it as a Mo settlement that Nafanas migrated to, and in time the Mo became disgusted and moved away (BRP 2002, 45). While I hesitate to assign these remains any sort of "ethnic" affiliation, Mo or otherwise (see Stahl 1991), varying reliance on staple grains may reflect seasonal or social differences and confirm the hybrid, frontier-like character of the Banda landscape at this time. Food may have even served as a means to come together over shared tastes in this multiethnic landscape. Pearl millet alone appears to have been desired in every village, since the grain occurs at all sites ranging from 20–100% ubiquity.

Yams (*Dioscorea* spp.) were probably also a critical part of peoples' diets in Banda, though we lack direct evidence since tubers do not preserve well. However given their mention in Bowdich's account of Asante, we know they were a desired food staple in the broader culinary landscape. Today, Banda lies within Ghana's most productive yam zone, and the desirable rainfall levels during the eighteenth and nineteenth centuries suggest an ideal habitat for yams. We can hypothesize that they comprised a major part of Banda's foodscape, though this awaits empirical verification.

Wild plants increased the variety of nutrients and tastes in what were likely starch-dominated diets. People may have eaten several leafy greens whose seeds were recovered, including *Cassia tora*, members of the Chenopodiaceae or Amaranthaceae families (cheno-ams), at least two *Portulaca* species, and *Zaleya pentandra*. As mentioned in the previous chapter, *Cassia tora* imparts a slippery texture to soups, suggesting a continued preference for this mouthfeel. People also enjoyed baobab, perhaps as an added leaf to soups, or in the form of a sweet-sour drink that is still made today out of baobab seeds. Wild plants were important for nonfood uses as well, as suggested by a cache of kapok seeds (*Ceiba pentandra*); kapok provides not only seeds that can be fermented into a flavoring agent for soups known as *kombotoo* (chapter 5), but also fiber used for bedding and pillows today. Shea butter was probably the main plant oil, providing families with an important source of fat and flavor, but it does not show up often or in great quantities in the archaeological record, perhaps hinting that oil was acquired through trade or produced outside the villages. Today shea butter trees are found in quantity only at the far northern edge of Banda territory, and the wetter conditions of the eighteenth and nineteenth centuries were probably even less ideal for the encouragement of local groves.

While people continued to eat many of the same plant foods, meat consumption is one area that was drastically different than in preceding centuries. Mean density of animal bones decreased significantly across all classes of animals, from rodents and lizards to cows and sheep. The mean density of bovids and artiodactyls, classes which include domestic sheep and goats as well as cattle, are at their lowest point in Banda's millennium-long occupation (Logan and Stahl 2017, 1388). Herders may have had a more difficult time keeping animals healthy, as wetter conditions would have increased the tsetse fly populations and increased the risk of sleeping sickness. Livestock was also in high demand in Asante's forest core (Arhin 1987, 57; Bowdich [1819] 1873, 272), where it was even more difficult to rear animals. After they were conquered in 1774, Banda's leaders agreed to provide Asante a yearly tribute in sheep (Wilks 1975, 246); by the late nineteenth century, Banda was on a well-known livestock trading route (chapter 4) that may hint at early nineteenth-century antecedents. We know less about how this apparently reduced meat consumption impacted health and cuisine.

That people supplemented their diets with wild animals that could be acquired locally seems clear. Animals like rodents and lizards that are easily acquired through opportunistic encounters are present in the record. Overall, however, people acquired a much less diverse array of animals than in previous centuries. The dangerous animals that would have been encountered at the markets of Kuulo Kataa are virtually absent from Makala Kataa (Logan and Stahl 2017; Stahl 1999b). This pattern may hint at an absence of skilled hunters, which makes sense given the influx of Nafana newcomers who would have had to acclimatize to their new landscape. Some protein tastes were acquired from the Asante heartland (McCaskie 1995, 29), particularly for snails (Stahl 1999b, 54). But compared to the Asantehene's diet, that of the average Banda resident was probably quite modest. In particular, there is less meat from a much narrower range of animals.

Food preparation can tell us a lot about status and local tastes, although at present our available archaeological data speak to this issue only in a limited way. Phytoliths (see chapter 1 and appendix A for description of method) are microscopic silicified plant tissues that are found on the tools people used to prepare food as well as in sediments in areas where people prepared and consumed food. While this type of analysis has not yet realized its potential in the African continent, a small sample of artifacts and sediments were tested from Banda. An earlier study by Deborah Pearsall (see Stahl 1999b, 35–36) found phytoliths that are used to identify maize in South America at Mounds 5 and 6 at Makala Kataa.[9] Subsequent methodological assessment suggests that these phytolith forms may derive from wild and domesticated African grasses and must be interpreted with caution (Logan 2012). However, using a refined methodology, I analyzed four soil samples and residues from two artifacts (see sampling details in appendix A). Results generally confirm the pattern observed in charred remains: probable presence of pearl millet in every context sampled, with many also containing sorghum, and no strong indication of maize (appendix B; Logan 2012).

Two contexts are worth special mention. The first is an area immediately below a grindstone in the MK Mound 6 structure (2W 24N, Lev. 8) which contains a very high proportion of sorghum phytoliths, including several types that may be unique to sorghum. There is also a moderate probability that sorghum is present on a spherical hand grinder recovered from the MK Mound 5 kitchen area (0W 2S). These data hint that sorghum was being processed into a ground food product, contrary to Nafana oral histories which insist it was not eaten.

Another way we can assess food preparation is to look at the spatial organization of food activities. An exceptionally well-preserved kitchen was uncovered in Mound 5 at the village of Makala Kataa. The kitchen was probably used between 1774 and the 1820s, after which it was rapidly abandoned, leaving many of the original artifacts and foodstuffs in place (Stahl 2001, 169–71).[10] While most of the other archaeological contexts I discuss were accumulated in diverse ways and often over long time periods, the Makala Kataa Mound 5 kitchen provides rare insight into food preparation. In what follows, I attempt to transport the reader back in time to this kitchen to better understand the experience of its inhabitants. The following narrative should be paired with figures 2 and 3, which show the spatial distribution of these activities, as well as quantitative data on plant representation in appendix B.

It is a late afternoon sometime around 1825, and you are visiting the compound now buried in Mound 5 (Stahl 2001, 169–71). You are drawn to the focal point of the courtyard: the hearth, where the family women are busy preparing dinner. The cook looks up and greets you from one of the two hearths—a roofed one if it is raining (excavation unit 4W 4S), or the three-stone laterite hearth nearby (2W 4S) if the sky is clear. She is preparing a sauce of wild greens, selected from nearby pots (4W 4S) which may contain the herbaceous *Portulaca*, a favorite throughout West Africa, or one of the many edible species belonging to the Chenopodiaceae or Amaranthaceae (cheno-am) families.[11] Fish (perhaps dried) stands ready in a pot (2W 4S) nearby to be added to the soup. Carefully hung in the rafters and stacked beneath the pole-and-thatch roof of the wet-season kitchen are piles of harvested sorghum and pearl millet heads and stacks of pots containing ingredients (4W 4S, 4W 2S).[12] Just a couple meters away (2W 4S, 0W 4S), grains are being liberated from glumes and stalks as part of their transformation into food.[13] Some of the older daughters are busy processing grains and grinding them on the large grinding stones that ring the fire (0W 4S, 6W 0S to 2E 2S), after which they will be cooked and beaten into a thick porridge.

While some of the women are cooking, other household members are engaged in other activities or are simply relaxing in the compound. A few older men and women are smoking a pipe full of tobacco near the bank of rooms to the northeast (0W 0S to 4E 2S). The cook too will get her turn: a pot with tobacco sits ready in the wet-season kitchen area (4W 2S). Had you arrived somewhat earlier one of the women would have been busy crafting pots, creating just the right paste

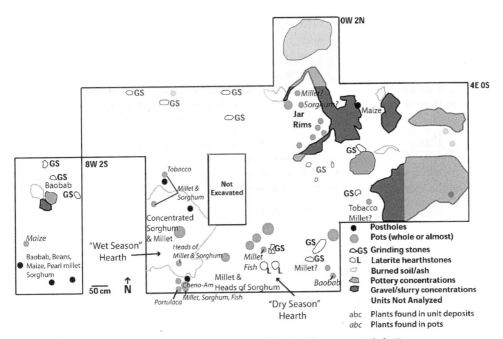

FIGURE 2. Plant remains identified in Early Makala kitchen deposits at Makala Kataa, Mound 5 (base map redrawn from Stahl 2001, 170).

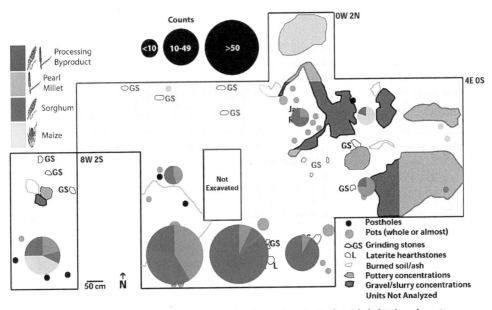

FIGURE 3. Distribution of grain crops and their byproducts in Early Makala kitchen deposits at Makala Kataa, Mound 5.

mixture on grinding stones (8W 2S), or using specially fashioned convex sherds to smooth and thin pot walls and a maize cob to decorate their outside, stowing these tools in a pot (8W 4S) after finishing for the day.[14] Another person nearby is busy processing foodstuffs including baobab, beans, maize, pearl millet, and sorghum, under a pole-and-thatch roof (8W 4S). The younger children are already hungry and have managed to hunt down some wild figs (*Ficus sp.*), spitting out their seeds nearby (0W 0S, 0W 4S, 4W 4S).

Then the worst happens. Whether because of an accident or because warriors from a neighboring polity suddenly exploded into the village, this tranquil scene is enveloped in fire. It probably started in a cooking hearth, turning to ash most of the ingredients ready to be included in the day's main meal, and charring the carefully harvested piles of sorghum and pearl millet.[15] In a panic, people dropped everything—pipes, pots, and food—and ran to safety, perhaps to the large cave in the Banda hills that had sheltered their grandparents fifty or so years earlier during the Asante invasion. An enterprising individual grabbed the full storage pots or calabashes that had been nestled in the carefully arranged jar rims that served as pot stands in the storage area (0W 0S); the lack of preserved plant remains and other heavily burnt materials suggest this area escaped the fierce blaze that consumed the southern part of the compound. Whatever the cause of this sudden abandonment, household members left behind most of their kitchen equipment and stored grains, and were not able to fetch them later.

TASTING FEAR IN THE MIDST OF VIOLENCE

As the rapid abandonment of Makala Kataa suggests, the nineteenth century in the Volta basin and throughout large swathes of West Africa was a period of widespread violence and dislocation as groups vied for power and battles erupted (Arhin and Ki-Zerbo 1989; Law 1995). This was related in part to the instability of states like Asante, which faced increasing military and economic pressure from the British beginning in the 1820s. The abolition of the slave trade in 1807 dealt a major blow to Asante economic power, given the central role of the slave trade in Gold Coast trade at the time (Arhin 1990; Arhin and KiZerbo 1989, 687). As centralized power weakened, so too did Asante's authority over the margins of its empire, leaving its outer territories vulnerable (Arhin and Ki-Zerbo 1989, 663–71; Wilks 1975).

A major source of unrest in areas to the north of the forested region was slave raiding for the internal market, which, despite the British abolition of the slave trade in 1807, continued into the early twentieth century in some places. This was related to the strong domestic demand for slaves as tribute payments or as laborers, in part to fuel the increasing production of cash crops and gold mines, and as payment for firearms (Arhin 1967b, 76, Arhin 1977, 4; Arhin and KiZerbo 1989, 686; Law 1995). Slaves were captured either through slave catching (i.e., kidnapping) missions or as a byproduct of violent conflict. Decentralized or stateless societies were often the target of slave catching and raiding, but they too had developed strategies for dealing with this threat. Though the nineteenth century saw intensification in slaving activities over much of the Volta basin, even in earlier times stronger polities had regularly demanded captives from weaker ones, especially to meet tribute obligations (i.e., Asante demands from Gonja and Dagomba). To avoid capture, people could migrate, develop defensive settlements and strategies, and/or participate in the trade by capturing others (Swanepoel 2005, 274–78). Migratory movements were very common in the nineteenth century, and in many places they underwrite the present-day map of the distribution of ethnic groups (Arhin and Ki-Zerbo 1989, 662; Whatley and Gillezeau 2011).

The pressures brought on by the slave trade and the new ways of living it engendered had multiple impacts on food security. Based on historical sources, Cordell (2003, 40–42) outlined the subsistence strategies of early nineteenth-century central African societies. He points out that monocrop grain fields were particularly vulnerable during raiding: they were not only highly visible as distinct from nearby vegetation, but also required the presence of a farmer at very specific times of year for care and harvest. One adaptation, recounted in numerous historical sources, was to rely more strongly on hunting and gathering, which would leave few signs to be read by hostile marauders. Another possibility was to switch to the cultivation of cassava, which blended in with the surrounding vegetation, required little care, and could be left in the ground for several years and be harvested at any time of year (see also Ferme 2001; Stahl 2008b, 40).

Oral histories of the Banda region recorded at numerous points (Ameyaw 1965; Owusuh 1976; Stahl and Anane 2011; Stahl 2001) emphasize multiple migrations and conflicts and are illustrative of many peoples' experiences during the turbulent nineteenth century. The major period of dislocation seems to have begun after the Banda people joined Asante in war against the neighboring polity of Gyaman (map 3) in 1818–19, eventually fleeing to nearby Bona (Ameyaw 1965). Quarrels and eventually conflict resulted, at which point the people of Banda sought refuge in Gyaman, and with its aid launched retaliation against Bona. Asante entreated the Banda peoples to return to their present-day location, but Gyaman soon attacked, leading Banda peoples to flee the area, this time moving to Mo/Nkoranza (map 3) territory north of the Black Volta River. Disagreements once again erupted with their hosts, and Banda peoples dispersed to various locations, including Bui (chapter 4), a town just south of the Black Volta. Banda again joined the Asante in an 1893 war against the Nkoranza, and the Nkoranza in turn terrorized Banda villages (Arhin 1974, 113; Stahl 2001, 190). Under the new chief Sie Yaw, Banda peoples eventually recongregated at Bui and near modern-day Banda-Ahenkro, after regaining possession of their land, which had fallen into the hands of Gyaman. In other words, over the course of about one long lifetime, Banda peoples experienced multiple violent conflicts and long-distance moves, which must have had major impacts on their livelihoods and sense of security.

Oral historical sources characterize these times in tragic (perhaps exaggerated) terms (Ameyaw 1965 in Stahl 2001, 155). In the oral histories recounted to me by elders in 2009 (chapter 4), based on the experiences of their grandparents and great-grandparents, people recalled that food was in short supply, and people were often too mobile to farm. Most individuals who remembered the stories passed down from their older relatives tell tales of starvation and people left behind (see also Stahl and Anane 2011). There is a sense of haste and motion in the stories, of running and quick decisions, of people and places abandoned in flight. While this highly mobile state surely did not characterize the entirety of the decades-long gap in settlement that we see archaeologically, the moments of flight were traumatic enough to become deeply etched onto cultural memory.

On the move, people hunted wild animals and ate leaves from the bush, along with whatever they were able to grab as they fled their homes or could take from farms encountered in their journeys. Interestingly, maize, millet, sorghum, and yams are not mentioned. This may not be accidental. All of these crops would have required a certain amount of seasonal maintenance, and probably would have withered quickly once their human caretakers departed. Plucking cassava tubers from abandoned farms is mentioned, however, which also makes sense as this is a low-input, hardy crop that can be grown in almost any conditions and at any time of year (see below; cf. Cordell 2003; Ohadike 1981). Wild leaves are an important part of the narrative about the wars; nearly every person who recalled oral histories referring to these traumatic times mentioned wild leaves (chapter 5).

One place on the landscape figures prominently in oral histories of nineteenth-century turbulence: Banda Rockshelter, which was repeatedly used as a refuge in times of violence. People fled there during Asante's pursuit in the dry season of 1773–74 (Stahl 2001, 157–58), as well as in later conflicts, such as with Gyaman after 1819.[16] Food shortage is explicitly referred to in both contexts. Famine reportedly forced those hiding in the rockshelter to surrender to Asante (Ameyaw 1965 in Stahl and Cruz 1998, 214), and oral histories stress a return to hunting and gathering during the mid- to late-nineteenth century wars. Though it is impossible to determine the specific conflict to which the material remains in the rockshelter belong, they can speak to activities and strategies adopted during periods of distress.

Unlike village sites, the rockshelter was not excavated extensively, in part due to its smaller areal extent.[17] Food storage and preparation containers are present. The only other artifacts are a single ceramic bead, a locally manufactured pipe, and porcupine quills (BRP 2002). Botanical remains were likewise scarce in the nine samples available for analysis; even charcoal was present only in limited amounts. Two maize cupules are present, as is one cowpea, and one possible grain of pearl millet. Several different wild seed taxa were discovered, though most could not be identified due to their fragmentary and distorted nature (n = 4 types). Two types were provisionally identified as Apocynaceae and Malvaceae, and appear to be novel to this phase, as neither of these taxa were recovered from samples dating to earlier phases (appendix B).

Though these data are meager, they conjure an image of privation in a dark dank place where use of fire was probably minimal for fear of being discovered. The presence of maize, pearl millet, and cowpea is consistent with diets in Early Makala times, but there are several important differences. Sorghum is absent, but maize, rare during Early Makala times, is present. Maize's quick-growing qualities may well have been advantageous during periods of hiding. Though impossible to trace, it seems plausible that cassava was also eaten at this time, especially given its prominence in oral historical accounts and historical documentation in other parts of Ghana, which speak to its widespread use as a famine food (La Fleur 2012). Finally, although it is not possible to narrow down which wild foods were eaten, they clearly differ from foods consumed during the heyday of the Early Makala occupation of Makala Kataa, supporting oral historical accounts of experimentation.

Banda was not alone in experiencing food insecurity during these decades in the middle of the nineteenth century (see chapter 4). Widespread violence and dislocation characterized much of West Africa at the time, and violence is among the strongest predictors of severe food insecurity. This was a crisis with both economic and political roots. When the Atlantic slave trade was outlawed by the British in 1807 a prime source of wealth was removed from African polities like Asante, which turned to "legitimate trade" items like gold and oil palm. Yet those commodities also demanded considerable labor inputs, which led to an increase in the internal slave trades (Law 1995). At the same time, the power of Asante was waning as the British struggled to find a foothold on the coast and engaged in repeated conflicts with it. Thus distracted, Asante's control was weakened over marginal areas like Banda (Arhin and Ki-Zerbo 1989).

It is difficult to find information on the nutritional impacts of this widespread violence, but I suspect they were significant in the regions directly affected. Austin, Baten, and van Leeuwen (2012) illustrate a significant decline in the heights of men born in the 1840s from northern Ghana and Burkina Faso, areas that were at the epicenter of mid-nineteenth-century violence as Europeans struggled to gain control of interior West Africa. Height is both genetically and nutritionally determined, with protein consumption being the strongest predictor of overall adult height. Interestingly, there are no significant height declines in cohorts born in the 1800s–1830s, suggesting that there was a significant interruption in nutrition in the 1840s but not before that, at least among the populations for whom we have records. Given the association of meat with social status in Asante's core, this sharp drop in protein consumption may well signal sharp increases in inequality as well.

TASTING CAPTIVITY

Violence splintered social groups and created new ones throughout the period of the trans-Atlantic slave trade and legitimate trade (Whatley and Gillezeau 2011). Victors reaped the spoils of war and left triumphant, while the defeated paid the costs. For some, the costs were especially high. Labor was in high demand internally and from European slave traders along the coast, and the Asante actively acquired human beings to be consumed as commodities in both trades. Banda was no exception. A 1774 missive by Pieter Woortman, the Dutch director-general of Elmina, reports better than usual trade in slaves thanks to the Asante defeat of Benda [Banda], which had sent him a "considerable number" of slaves (Yarak 1979, 58; Stahl 2015, 267). These captives may have encountered famine and starvation once they arrived on the coast, as multiple sources record food shortages in that same year (La Fleur 2012, 141). Although we do not know where the newly enslaved from Banda were destined for, nor how many souls were sold, the documentary record reveals the scale of the slave trade, as well as some of the horrors of experiences of captivity at Elmina and in the ensuing Middle Passage. Food, and especially maize, was central to the deprivations suffered during these travails.

Historians, geographers, and economists have long been convinced that there is a relationship between American crops and the slave trade. "As for the influence of these crops before 1850," Crosby ([1972] 2003, 168) proposed, "we might hypothesize that the increased food production enabled the slave trade to go on as long as it did without pumping the black well of Africa dry." Alpern (1992, 13) notes that "these crops are said to have improved diets and accelerated population growth, to the point, some would argue, that human losses through the slave trade were more than offset by the enhanced ability to feed people" (see also Curtin 1969). Although this idea has been critiqued by demonstrating the stagnation in population growth in the eighteenth and nineteenth centuries (e.g., Rodney 1972), in modified form it continues to be repeated and tested in recent scholarship in economic history (e.g., Austin, Baten, and Bas van Leeuwen 2012, 1299; Cherniwchan and Moreno-Cruz 2019). Referring to this thesis as the Crosby-Curtin Hypothesis, Cherniwchan and Moreno-Cruz (2019) test the idea that where maize spread populations increased and the trans-Atlantic slave trade peaked. Lacking sufficient data on the precolonial spread of maize, they map out environments suitable to maize cultivation in modern times to substitute for the lack of ancient data. Their country-scale analysis suggests that countries that are best suited to cultivate maize ecologically saw both population increases and greater engagement in the slave trade. They interpret this correlation as indicating that maize resulted in increased populations and thus more people to trade.

Arguments like these are highly problematic for both empirical and conceptual reasons. Estimating the spread of maize based on modern-day ecological data is erroneous since we know environmental conditions were much different at the time of maize's adoption (Shanahan et al. 2009; chapter 2). This kind of estimation also presupposes people automatically adopted this cultigen, removing agency in this decision-making process. As we saw in chapter 2, maize adoption was very slow even in Banda, which is ecologically well suited to its cultivation. And Banda was not alone in slowly adopting maize, as emerging archaeological studies show (Gijanto and Walshaw 2014; Esterhuysen and Hardwick 2017; Widgren et al. 2016). In sum, as with earlier scholarship on this topic, the data used to estimate maize adoption by Cherniwchan and Moreno-Cruz is circumstantial evidence at best;

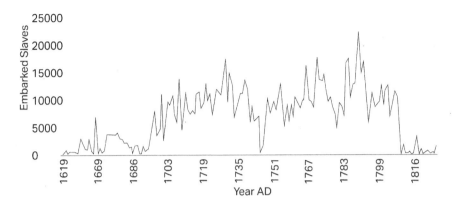

FIGURE 4. Numbers of slaves embarked yearly from the Gold Coast, 1619–1840. Note scale change on X axis between 1619 and 1669. Data from transatlanticslavevoyages.org.

the limited archaeological studies of maize's spread suggest that their model of spread is also empirically incorrect. Archaeological case studies of maize adoption across the diverse ecological and cultural landscape of the continent are sorely needed to understand agency and contingency in the process of food adoption.

If American crops did help make up for population losses suffered to slavery, we would expect there to be a correlation between maize adoption as a staple and the height of slave raiding and exports; indeed, this is precisely what Cherniwchan and Moreno-Cruz set out to test. The time frame encompassed in this chapter corresponds to the height of the Atlantic slave trade in Ghana (figure 4; see TransAtlanticSlaveVoyages.org). From 1700 to 1830, over one million captives were forcefully taken from the slave ports along the Gold Coast. Some of Banda's people were among them. Yet in Banda, an area we know suffered population losses, there is no substantial increase in maize consumption during this time frame, with the exception of a perhaps temporary increase at Banda Rock-shelter. Available data from the rest of Asante also do not support an increase in everyday maize consumption among the populace. Additional archaeological data from areas on the margins that were the hardest hit by slave raiding are needed to evaluate this hypothesis. It is only people who inhabited the coast and its hinterlands who appear to have adopted maize as a staple during this time. There, cooks experimented with maize and developed new foods like *kenkey*, a fermented maize dough that is widely consumed today. What were the circumstances there, and how were they different than in the Asante core and its margins?

Our resolution on coastal foodways is much finer because European engagements there occurred earlier and were more regular, hence the documentary record is more complete. Recent Africanist food scholarship has suggested that maize played a critical role in provisioning captives as well as slave transport ships. Maize had long been provided to captives along the coast, first by the Portuguese, who considered the crop unfit for their own consumption (La Fleur 2012, 92). The association of maize and the enslaved seems to have continued into later centuries, as the Gold Coast itself became a major exporter of human cargo especially after 1700 (Lovejoy 2011). Maize provided an economic opportunity for coastal farmers and traders who could easily turn a profit trading the grain to slavers and ship captains (Carney and Rosomoff 2009; Miracle 1965). High demand for foodstuffs elevated crops from significance in local and regional trade to significance in international commerce—the cash crops of the early modern era. In the long run, a shift to the exporting of human captives must have had major impacts on agricultural production, since it depleted the rural labor force of men and women in their primes (Carney and Rosomoff 2009, 46–56). Yet at least along the coast, we see an intensification of agricultural production through maize to take advantage of the demand for grains from European ships. We can only speculate as to whether enslaved labor was used to cultivate maize along the coast. While providing economic security for local farmers, maize was also purportedly the staple of slave transport ships (Miracle 1965), and thus "a symbol of the dehumanizing condition of chattel slaves, who were no longer able to exercise dietary preferences or choose the type or amount of food they consumed" (Carney and Rosomoff 2009, 55). Maize, in this formulation, falls from being the "grace" or savior of Africa (McCann 2005) to being a reminder of one of the darkest periods of its history. "This Columbian Exchange crop flourished because its cultivation and preparation made the cereal ideally suited to the trade in human beings" (Carney and Rosomoff 2009, 55), not because the forest lacked in carbohydrates.

FOOD AND INEQUALITY IN ASANTE AND ITS MARGINS

For the last half century, discussion of Asante foodways has focused on food supply and the potential role that American crops played in increasing food availability. In this chapter, I have tried to debunk some of the assumptions of the Big Bang hypothesis, and the related idea that American crops made up for population losses suffered during the slave trade. Similar starting assumptions are at the root of both explanations: that the context into which these new crops were received was lacking in adequate food supplies and local food traditions, and that Africans did not possess the knowledge to develop solutions through exploiting local environments. In this chapter I have offered a different and more complex narrative by highlighting some of the diverse African agencies and situations that characterized the eighteenth- and nineteenth-century foodscape. This approach brings to life the contexts into which American crops were received, rejected, and adopted, a diversity that troubles the foundations of simplistic Malthusian explanations for their adoption.

Foodscapes of the eighteenth to early nineteenth centuries occupied the full spectrum of choice and possibility that map on to increasingly unequal power relations. While I have only touched on the role that food played in status displays of the Asantehene and his retinue, the wide range and quantity of meat and plant foods consumed by elites were impressive. They resulted from a highly organized, productive agricultural system built on yams, plantains, and multiple kinds of grain, as well as an extensive empire that extracted food and many other products from subject nations like Banda. American crops seem to have been absent from these sumptuous displays. This absence is notable since other exotic foods were clearly desirable, hinting that American crops were not part of distinction-making practices. The continued preference for local foods is in stark contrast to the desire for other foreign consumables, chief among them alcohol. Local alternatives like palm wine were replaced with foreign rum and schnapps whenever possible, and these beverages continue to play important roles in marking power among chiefs even today (Akyeampong 1996; for cloth and tobacco, see Stahl 2002).

Yet many of the Asantehene's subjects lacked access to this full range of foods and may have preferred other kinds of foods that were valued locally. The people who lived in Banda subsisted primarily on pearl millet and sorghum (and likely yams as well), which had been the staples of their ancestors. Maize remained a minor component of their diets, even if it was more widely available than it had been in the past. People's meat consumption changed from preceding time periods; not only do we find a smaller quantity of meat, but people focused on more easily acquired animals rather than the novel, diverse menu that their ancestors had enjoyed. While there are many possible explanations for this change, it is clear that the Asante Empire had a major impact on local political economies. Banda's long-distance trade networks seem to have contracted and been redirected towards Asante. Banda's residents were also obliged to pay tribute to Kumasi in the form of livestock.

While Asante control impacted foodscapes and daily life for subjects on the margins, its most devastating effects were felt during periods of violence and its aftermath. We know of at least two periods of violence in Banda, associated with Asante's takeover in 1773–74 and with wars beginning in the 1820s and continuing periodically through the mid-nineteenth century. Oral histories record people fleeing to take shelter in the Banda hills and subsisting off what little produce they could acquire as well as leaves and animals from the bush. Maize and cassava were important in conditions like these in Banda and elsewhere along the Gold Coast, when culinary preference was secondary to simply meeting dietary needs. For the areas hardest hit by violence, the disruption in normal food supplies seems to have impacted lifelong growth and development, as observed in a drastic decline in height in men from northern Ghana and Burkina Faso in the 1840s.

This brief survey of foodscapes of the eighteenth and nineteenth centuries reveals that American crops such as maize were not adopted because African agriculture was incapable of supporting Africans. Where people had access to pre-ferred foods, American crops were usually not what was chosen. Where it made economic sense, such as along the coast, farmers and traders made a strategic choice to grow, eat, and trade maize. When choices were constrained during bouts of violence, people made the strategic choice to cultivate and consume maize and cassava, the crops that grew the quickest or easiest. And when choice was denied altogether, people ate what was provided them or made the agonizing decision not to eat at all. In the next chapter, I show how this diverse and unequal political and culinary landscape contributed to the making of chronic food insecurity in the generations to come.

Creating Chronic Food Insecurity in the Gold Coast Colony

In the nineteenth century, considerable turbulence was unleashed along the vast trade networks that had for so many centuries defined social and economic interactions from the West African coast to the Sahara. This turbulence resulted from a complex array of sources. The Atlantic slave trade was formally brought to a close, cutting off many polities from a prime source of income. At the same time, economic recession, political maneuvering, and ideological shifts across the globe led European governments to scramble for African land and resources after the trade in African people was outlawed. The friction among these desires contributed to local hostilities, resulting in violence and dislocation over wide swathes of regions encompassed by the modern nation-state of Ghana and beyond. By the end of the nineteenth century, the imposition of colonial boundaries had quelled much of this upheaval, but for many areas did little to make up for the losses already suffered. Colonial governments ushered in a new era, one in which the value of commodities was defined in European capitals, which in turn etched "development" unevenly across African landscapes and social groups.

The impacts of these economic and political shifts in the late nineteenth through mid-twentieth centuries varied across the landscape. In some areas, these changes provided the necessary preconditions for the entrenchment of chronic food insecurity. Demographic collapse and the imposition of market-based economic policies fixed the seasonal food shortage known as the hungry season gap into a permanent feature of some African livelihoods. Yet these forces remained largely invisible to British officials, who tended to suffer from historical amnesia and often viewed Africans as lazy and backwards. In this chapter, I follow in the footsteps of numerous scholars who have demonstrated the invalidity of these timeless stereotypes. Instead, I investigate how Africans coped with the loss of human capital—both people and the knowledges and labor they provided—and remade their lives in the uncertain conditions that prevailed in the colonial period. Returning again to Banda, I show how early colonial (1890s–1920s) policies indirectly impacted peoples' ability to feed themselves, and the ways that farmers and cooks maneuvered their limited resources to make ends meet. I also visit the Tallensi, situated to the far north of Banda, to survey how later colonial (1930s–1940s) attempts to understand African peoples and nutrition scientifically did little to ameliorate experiences of hunger.

COLONIALISM AND FOOD SECURITY

In the historiography of Africa, the impact of colonialism on African health and well-being has been a long-standing if unresolved research question (Lappé and Collins 1978; Worboys 1988). Colonial economic policies, particularly the focus on cash-cropping and market economies, fundamentally altered peoples' relationships with the land and its products (Giblin 1992; Doyle 2006; Iliffe 1987; Mandala 2005; Moore and Vaughan 1994; Watts 2013). Whether those impacts were negative or positive has long been debated. Earlier scholars tended to assume "that in pre-colonial times hunger, even famine, was more prevalent than today, despite the dramatic droughts of the Sahel region and the rapid growth of population in recent times" (Goody 1982, 58). The idea that American crops made up for losses suffered to slavery (chapter 4) relies on a similar set of presumptions: that Africans were worse off in precolonial times, and that European interventions increased food security. Assumptions like these were rarely evaluated by means of evidence, but instead appear to be the product of the ideologies common at the time, as well as of the use of a baseline precolonial Africa derived from early colonial referents. In this section, I train attention on how the impact of colonialism is assessed, specifically focusing on how scholars have dealt with the so-called precolonial period.

One of the main challenges to evaluating the impact of colonial interventions is the limited information—or perception thereof—that we have about precolonial food security (Doyle 2006, 1; Shaffer 2017). As I suggest in chapter 2, this is not necessarily due to a limitation of the archive per se, but rather a lack of dedicated focus to evaluating precolonial food security. People not specialized in early African history and archaeology often fall back on a imagined baseline precolonial period for comparative purposes, much as scholars like Goody and many that came before and after did.[1] Audrey Richards, for example, used the start of her ethnographic mission in the early 1930s as her referent for "traditional" Bemba, glossing over the significant changes of the preceding decades (Moore and Vaughan 1994). In the last several decades, contemporary scholars have often used a late nineteenth-century referent, a time after the ravages of the trans-Atlantic slave trade and the troubled decades of violence in the mid- to late nineteenth century. This point is chosen not because it is most representative of precolonial history, but because the documentary record expands considerably in this period, making it more visible and accessible, particularly to the nonspecialist. However, as we have learned in the previous chapter and will continue to detail below, these decades represent a low point in historical food security in West Africa.

A case from northern Ghana illustrates some of the problems with this kind of approach. Shaffer (2017) pulls together information from an impressive array of disciplines, including linguistics, history, and ethnography, to evaluate whether colonial policies had a negative or positive impact on local food security in northern Ghana. He argues that hunger characterized the precolonial period, based on the existence of words for "hunger" in local languages recorded by Europeans in 1899 and 1906—observations made on the heels of extremely severe disruptions, detailed below—as well as references to hunger recorded in twentieth and twenty-first century greetings and proverbs. As Stephens (2018b, 791) has outlined, it is tremendously difficult to date proverbs, so Shaffer's use of proverbs to imply precolonial food insecurity needs to be interpreted with extreme caution. The precolonial period is also collapsed into what two Europeans could record on their brief visits in 1899 and 1906—which actually falls within the colonial period, and is certainly not representative of food security in the vast "precolonial" period. Although Shaffer (2017) acknowledges that these are weak sources, he nevertheless argues that the hunger season existed before colonial rule, despite not identifying a single source that alludes to this specific type of seasonal food insecurity.[2]

But whether or not hunger existed before colonial rule should not be our focal point: of course hunger occurred periodically, as Watts (2013) details for northeastern Nigeria and James LaFleur records in coastal Ghana (La Fleur 2012, 137–44), as it did elsewhere in the world. Banda's story suggests, however, that there was a much higher degree of resilience to drought and other sociopolitical troubles in precolonial times than has previously been understood. This case study prompts a different but related question, about whether the form and severity of hunger changed during the colonial period and thereafter. John Iliffe's (1987) meticulous study of the African poor across the continent argues that famine in particular decreased during the colonial period, but that low-level, more chronic poverty (and presumably, food insecurity) increased during this time. Famines are severe, rare events which often lead to considerable deaths from starvation and disease, and because of their severity tend to be remembered across oral and written archives. Often portrayed as resulting from severe environmental changes like drought, famines more typically result from a failure to distribute food supplies to those who need them most (Sen 1981). Iliffe claimed that colonial governments improved food distribution networks through the introduction of modern transportation, a technological development which has been credited with famine reduction all over the world. Note, however, that the technological capability to distribute food does not mean there is always the political will to do so (e.g., Davis 2001). Even with this technological ability, famine continues to reoccur in African contexts. While this tendency has been explained in multiple ways (i.e., entitlement failure, Sen 1981), part of the answer lies in the pervasiveness of chronic, seasonal hunger. Low-level shortages act to make people more vulnerable to major perturbations like famine (Watts 2013). While famines are often recorded in oral and written archives, chronic food shortage is less visible, making it a kind of "silent violence" (Watts 2013). The other difficulty in evaluating Iliffe's thesis is that he does not clarify the precolonial baseline he uses to evaluate the impact of colonial policies. Including this level of detail is a major challenge in any kind of comparative, synthetic history; when it is not included, the resulting suppositions tend to get repeated in subsequent literature all the same. One of Iliffe's most salient points was that colonialism's impact was not the same everywhere: in some regions, colonial policies deepened existing inequalities while in others these policies created new disparities. This observation underscores the need for regional case studies on food security during the colonial transition.

Scholars focusing on local case studies are more familiar with the diversity of archives, and consequently have tended to argue that food security declined as a result of colonial economic policies and their social impacts. In Malawi and Nigeria, Mandala (2005) and Watts (2013) suggest that precolonial African polities adhered to the tenets of a moral economy that insured the populace was well fed or at least better fed than in recent decades, and that this "subsistence ethic" was eroded under the transition to market economies. Although we must be cautious in recreating a "Merrie Africa" imaginary (Hopkins 1973, 10) in place of the scarcity slot, it is important to highlight that social strategies like labor organization and food redistribution are among Africa's chief developments in agriculture (Guyer 1984; Stone, Netting, and Stone 1990). Although noticed long ago by anthropologists, these innovations tend to go unacknowledged in other contexts because they are less visible than the technological innovations that define improvements to Western agricultural practice (Berry 1993; Guyer 1984; Stone, Netting, and Stone 1990).

Beginning with the pioneering work of Audrey Richards ([1932] 1964, 151–57), Africanists have highlighted the important role that chiefs played in insuring that their constituents were fed. The ability of a chief to feed his people helped him gain both followers and status. Provisioning would have thus been an important strategy in societies where the accumulation of people and their skill sets was valued (Guyer and Belinga 1995). Other social strategies also exist for leveling out uneven food supplies, such as food sharing between households and family members (Mandala 2005). Social relations and the expectation of food redistribution can be thought of as a critical entitlement (Sen 1981) that insured food security in some places. However, we cannot assume that these instances, drawn from specific nineteenth- and twentieth-century African societies, characterized all ancient African societies; to do so is to collapse regional variation and time and change in much the same way as the "baseline" approaches critiqued above. What they provide is an alternative example of how food security is maintained through social contracts. While Shaffer (2017) claims that changes to the moral economy are unrelated to colonial intervention, authors like Watts (2013) and Mandala (2005) point to the indirect but devastating impact of cash-cropping, wage labor, and individual accumulation, which together fundamentally challenged the logic of communal approaches. In the cases they consider, the destruction of these social relations and reordering of the relations of production undermined the ability of the least advantaged people to cope with environmental change (Watts 2013; see also Davis 2001).

Colonial policies were rooted in ideologies that discouraged many officials from seeing the benefits of African social strategies. These ideologies are numerous and often contradictory (Berry 1993) but worth briefly reviewing not only because they dictated approaches to agriculture in their new African colonies, but also because they infused the minds of many of the chroniclers upon whom we rely for information. The colonial mission was predicated on the idea that the African continent was rich in resources to be mined, rather than the epitome of scarcity it occupies in Western minds today. In the early colonial period and before, many British officials operated under the myth of exuberance—that tropical environments were rich and fertile, but that Africans were too lazy to exploit them appropriately (Curtin 1964). Such ideas helped motivate the colonial enterprise; the resources of Africa were ripe for the taking, if Westerners were willing to provide—or coerce—the labor needed to mine them. This stereotype thrived on historical amnesia; many African farmers suffered from a dearth of labor, following centuries of ruinous slaving, which Rodney (1972) argued limited their productivity (see also Carney and Rosomoff 2009). Indeed this productivity gap remains the focus of economists and development practitioners even today, often with very little acknowledgment of the historical reasons that it exists. Increasing the productivity of African farmers became part of the civilizing mission, but only if those farmers focused on cash crops.

At the same time, "there was a romanticization of pre-European African society which included ideas of moral innocence, a respect for African bush-skills, and a generalized notion of the noble savage" (Neumann 1995, 151). With preconceived notions like these, it is no wonder that most early colonial officials concerned themselves little with subsistence; it was generally assumed that Africans could feed themselves (Robins 2018). This negligence may have allowed most Africans freedom to cultivate what made sense to them, particularly in areas like Banda that lay outside of the main cash-cropping zones. This no-policy policy may also have meant that food insecurity flew under the radar for colonial officials, especially in the case of the low-level and chronic food shortages that Watts (2013) aptly termed "silent violence." Only later, once nutrition science made malnutrition visible beginning in the 1930s, did colonial officials realize the devastating impacts of ignoring subsistence agriculture, a subject I return to below.

These colonial ideologies were different in some ways from the scarcity slot mentality described in the introduction, but they show some ancestral similarities.

They were based on an idea of Africa as a land of plenty rather than scarcity. Attitudes about the capabilities of Africans were conflicting: they were viewed on one hand as noble savages with skills to exploit their environments, and on the other, as lazy people unwilling to shoulder the burden of productivity. Each of these ideas was rooted in ahistorical space, with little acknowledgement of history or change, a practice also central to the modern reasoning of the scarcity slot.

In the first portion of this chapter, I show how the labor dynamics of the late nineteenth and early twentieth centuries were a major departure from earlier periods, and were misread by colonial officials as evidence of unproductivity. I focus on the early colonial period to 1930 in the British Gold Coast, a time frame for which we have complementary information from Banda that allows us to assess the impact of colonial rule there. In the last portion of this chapter, I consider the later colonial period (1930–57), when the new sciences of nutrition and anthropology allowed colonial officers to see and at times to ignore the extent of the hunger problem that their own policies may have caused.

SURVIVING THE TURBULENT TIDES
OF EMPIRE BUILDING

Early British interests in Africa were defined by commerce rather than by a desire to rule. But by the mid-nineteenth century, British concepts of race had changed, and a push towards dominance ensued, following similar developments in other European countries at the time (Bassil 2011; Lorimer 1978). In 1868, Britain and the Netherlands swapped some of their holdings to consolidate their respective territories along the Gold Coast. In the same time frame, the Asante state tried to establish formal control over several coastal polities with whom relations had long been contentious, and also began to invade territories protected by the British. The British responded by sacking Kumasi in 1874 and again in 1896. With the loss of her southern provinces and access to Atlantic trade, the Asante state was considerably weakened. Although the independent Asante state persisted for two more decades, several of its internal provinces and other factions revolted. By 1901, Asante (now known as Ashanti) was annexed to the Gold Coast colony, and the Northern Territories were annexed as a protectorate in 1902 (map 5; Gocking 2005, 37–47). The British were not alone in their conquest of African lands and peoples. European countries vied for land and power over much of Africa in the nineteenth century, resulting in formal claims to colonies at the 1884 Berlin Conference, where not a single African was present (Forster, Mommsen, and Robinson 1988).

The Gold Coast interior was of particular interest, for it lay at the intersection of competing British, French, and German claims. In 1892–94, the British dispatched George Ekem Ferguson to the far north to obtain treaties of friendship with the polities there (Arhin 1974, ix–x). Ferguson, born of Scottish and Fanti parents, was a civil servant trained in London in geology who had traveled widely throughout

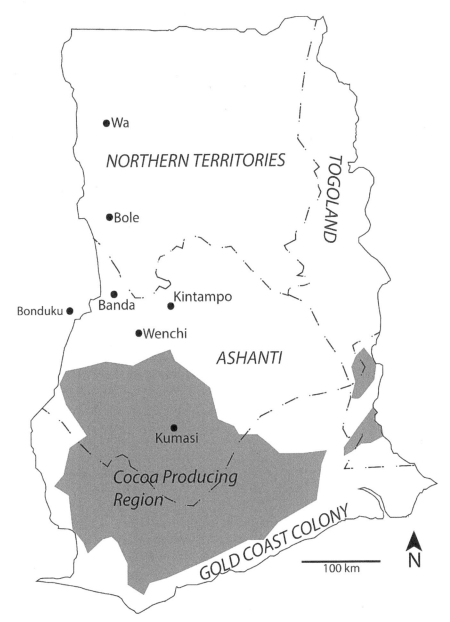

MAP 5. Divisions within the British Gold Coast, showing cocoa-producing regions, circa 1930. Note that the boundaries and names of regions have changed over time.

the Gold Coast in the service of various governors and later became a surveyor. However in the early and undeveloped colonial administration, Ferguson was employed as a jack-of-all-trades, particularly in settings that were considered dangerous for a European (Thomas 1972, 181–83). Ferguson apparently arrived on the heels of conflict in the Banda area, noting the "ruined villages of Banda" (Arhin 1974, 101, 113). The rulers of Banda signed a treaty with Ferguson in 1894, two years before Asante formally ceded to British control (Stahl 2001, 95–96). However, these initial British treaties were commercial in nature, and offered no guarantees of British protection (Arhin 1974, 66). This changed as British trade interests became threatened by the movement of Imam Samori and his Sofa army into the area (Stahl 2001, 96–97).

Samori and his forces waged wars of conquest between 1861 and 1898, and established a loose empire that stretched from Guinea to northern Côte d'Ivoire. By 1895, he had settled his base of operations in Bonduku, the center of Gyaman, some eighty kilometers west of Banda, and began expanding east and northwards into what is today Ghana (Person 1968). Samori succeeded in establishing a line of posts across the Asante hinterland including in Banda, Bole, and areas northwards. His armies brought great unrest, raiding the surrounding areas for food and captives and scorching much of what was left, causing widespread food shortage and dislocation (see below; Stahl 2001, 96–98). The greatest impact was felt north of the Black Volta River (see Fell's 1913 report in ARG 1/2/21/3; Stahl and Cruz 1998).

In January 1896 the British received word that Samori had occupied Banda; further investigation by a Gold Coast constabulary suggested that Samori had taken 112 captives, as well as foodstuffs, from Banda. Henderson suggested that five hundred troops were needed to retake Bole from Samori, though concern was expressed over how to supply so many men with food since "there is practically none between Lawra [Bui] and Buale [Bole] and there is practically none within a 15 mile radius of Buale [Bole]" (Henderson in Stahl 2001, 193). The British army established a base at Bui in Banda just south of the Black Volta River to staunch the flow of trade goods north to Samori. Their efforts eventually expelled Samori and brought peace to the area by 1898. After the dust settled, many towns were left ruined, and grain, cattle, sheep, and even wild animals were scarce (Stahl 2001, 96–98, 193–94).

How did people manage to get by during these turbulent, dangerous times? We can reconstruct some of their experiences through the archaeological record left behind at Bui Kataa, the remains of the village of Bui, occupied during and immediately after Samori's defeat. Bui is strategically located along the shores of the Black Volta River (introduction, map 2). Though the site is now submerged by the reservoir created by the recent construction of the Bui hydroelectric dam, archaeologists conducted rescue excavations in advance of the flooding (York 1965; BRP 2008). Due to time constraints these excavations were limited but they still provide glimpses into everyday life. People may have constructed homes to live in, as attested by the remainders of rectangular, beaten laterite floors uncovered in four mounds (York 1965).[3] That people invested in house-building suggests that they had begun to return to life as normal, an improvement over the privations attested to at Banda Rockshelter (chapter 3). Domestic activities resumed as well, as attested by cooking and storage vessels as well as iron implements. People seem to have retained or acquired some items of value too, including glass beads and cowrie shells, both of which may have been curated for personal adornment and ritual. Cowrie shells were also recovered and may have served as currency, a practice eventually replaced by British coinage. A handful of these items were imported from European sources, including ball clay pipes, white ware ceramics, glass beads, and substances contained in glass bottles, like liquor or medicine. But life in Bui was not without its dangers. Gun flints, shell casings, and bullets speak to the need to defend the village from Samori. Not everyone survived, as attested by one burial that showed injury consistent with a violent death (BRP 2008; York 1965, 18–19).

Banda peoples would have been tasked with feeding British soldiers as well as themselves. People altered what they were eating to cope with the increased pressure on diminished food supplies. One of the sacrifices that people made was adopting staple crops that grew more quickly. Maize became the staple grain, and the indigenous grains pearl millet and sorghum were abandoned, at least temporarily. Cassava, a low-labor crop that was considered poor man's food on the coast (LaFleur 2012) and in Kintampo (below), may have also been a dietary mainstay, as it was in other instances of dislocation (Cordell 2003; La Fleur 2012; as well as oral accounts I recorded in 2009). Notably, Ferguson mentions that his troops were unfamiliar with this crop when they encountered it in their 1894–95 tour of the Volta basin, a wide area that includes Banda (Stahl 2001, 208). A shift in dietary staples is a significant change in daily foodscapes, since tastes and textures of less-preferred foods are a constant reminder of vulnerability and lack of choice. Under different circumstances, we might interpret a shift like this as simply the emergence of new food preferences, but in this context of uncertainty it seems likely that maize and cassava indexed lack of choice. Maize's short growing season permitted two crops to be harvested per year, meaning people could produce more food quickly. Cassava can be grown with little labor, and its cultivation is easily hidden in the bush. After the devastations of Samori's pillaging, these qualities were likely needed at Bui, especially as the British soldiers stationed there placed extra demands on local food supplies.

Particular tastes, textures, and smells serve as sensory cues to past events and experiences, and familiar foods may have provided some comfort during these uncertain times. Although people had to rely on less-preferred starchy staples, cooks may have prepared them into familiar textures. Maize phytoliths were found on two grinding stones, suggesting that this less desirable grain was ground and perhaps made into a familiar food like *tuo zafi* or *koko* (porridge). The advantage of foods like *tuo zafi* is that they can be made from virtually any starchy flour into a culturally recognizable dish (chapter 5).

Although their choices were limited, cooks used familiar ingredients when possible. People seemed to have returned to eating familiar wild plants in their stews and soups, instead of the experimental variety attested to in oral histories and at Banda Rockshelter (chapter 3). These include *Ficus, Sida, Zaleya pentandra*, and cheno-ams, although any of these may represent weeds instead of food plants (Abbiw 1990). One of these (*Sida* sp.) would have imparted a slippery texture to soups, a mouthfeel that has a long history in Banda. Smoking retained importance, judging from the presence of tobacco seeds and pipes, and may have provided relief during stressful times. Peppers (*Capsicum*), another vegetable from the Western hemisphere, squash (Curcurbitaceae) and *Celtis integrifolia*, a common wild fruit, helped provide dietary variety and spice.

People's meat consumption also focused on expediency. Fish were consumed in higher proportions than are attested to in all of Banda's millennium-long history (Logan and Stahl 2017, 1381). From their location at river's edge, Bui's inhabitants were well positioned to take advantage of this critical resource. As an importance source of protein, fish would have helped to even out uncertain access to domesticated animals and crop plants, and fishing placed fewer demands on labor during peak agricultural seasons. People consumed other kinds of meat that would have been easily accessed in the immediate environs of their settlements and fields, particularly rodents and birds (Logan and Stahl 2017, 1381). The focus on these less-preferred animals may have been a result of overhunting by Samori's troops, but it also suggests that people were reluctant to stray too far from the village. Wild animals formed an important coping strategy during this troubled time, as evidenced by the high density of their bones (Logan and Stahl 2017, 1382).

COCOA NATION: CASH-CROPPING IN
THE BRITISH GOLD COAST, C. 1897–1930S

The British extended formal rule to the Gold Coast colony beginning in 1897, just after British troops had routed Samori's soldiers. They quickly set out to develop the economic resources of their new acquisition, particularly since it was expected to be self-financing, a policy Sara Berry (1984) termed "hegemony on a shoestring." A main thrust of their effort was in expanding trade in commodities already of interest, such as gold; a second thrust was the production of profitable export crops to fuel the machines and tastes of Europe. Starting in the early nineteenth century, the production of palm oil for export had been one of the earliest and most profitable readjustments to "legitimate trade" ushered in by the abolition of the slave trade. However, by the 1860s, alternate sources of oil had been found for European cosmetics and industrial applications and caused a depression in the price of palm oil. Rubber served as a substitute export crop in the 1880s and 1890s, followed briefly by coffee, but cocoa became the dominant export crop in the early twentieth century (Grier 1981, 32).

Cocoa exports grew very rapidly from their meager beginnings in 1891, when exports were worth just £4, to account for half of the colony's exports by 1911, with Ghana becoming the world's largest producer by 1919 (Gocking 2005, 47; Grier 1981, 32). The rapid expansion of cocoa production was to have dramatic implications for the nature of land tenure, labor migration, and the geography of economic development in Ghana (Hill 1963). The push for cocoa production was so great that many farmers in southern Ghana reduced or abandoned the cultivation of food crops, so that they came to rely on imported European foods in the early twentieth century. This was to have major implications when cocoa prices declined, leaving smallholder farmers open to considerable risk (Grier 1981, 32–34). Areas north of the cocoa belt, like Banda, continued to focus on subsistence crops, according them perhaps a measure of resilience, though they faced their own set of challenges.

By 1900, cocoa production had expanded to such a degree that acute labor shortages resulted. The colonial government found a solution by declaring the newly annexed Northern Territories—the land north of the Black Volta River—a labor reserve for both Ashanti and the Gold Coast colony (map 5). The resulting outmigration of able-bodied men to work in cocoa farms and gold fields in the south diminished agricultural production in the north. The rise of this migrant labor force also helped facilitate the penetration of the Northern Territories with goods from the metropole, another central goal of the British colonial administration. This had the dual if not altogether intended effect of eroding the self-sufficiency of subsistence farmers, thus making it necessary for them to earn wages through migrant labor to support their families in the north. The import of British manufactured goods also led to the collapse of many local industries, such as iron, ceramic, and cloth production (Grier 1981, 24, 37–38).

As part of the development of markets for British goods, colonial officers sought to monetize their colony under a uniform currency. Adoption of a new British currency instead of local cowries would presumably pave the way for the acquisition of British goods. One way in which this was accomplished was the requirement that taxes, fees, and fines be paid in British currency. While the adoption and accumulation of such currencies were slow and uneven across the territory, they had major impacts on socioeconomic organization and agricultural production (Guyer 2004; for Banda, see Stahl 2001, 99–101). One of the simplest ways for rural people to gain access to cash was through the sale of local products, particularly surplus agricultural products or cash crops and craft goods like pottery (below; Stahl and Cruz 1998). Over the long term, imported goods transitioned from the realm of desired goods to necessities. The selling off of agricultural surplus, instead of storing extra grain for use later in the year, probably played a major role in creating the hungry season gap that is now so widespread throughout the continent, a thread I explore through Banda's experience below.

While the Gold Coast colony and Ashanti were undergoing rapid development, particularly in transport networks, little effort was expended on extension of transportation networks to the northern half of the country. Though colonial officers experimented with the production of cotton, shea nuts, and other locally grown commodities in the north, most of these ventures failed because of the high transport costs to the coast. A visit by W. Tudhope, the director of agriculture, to the Northern Territories in 1912 recorded local production of cotton, and determined that the price that growers obtained locally could not be matched if cotton were exported to London. Though alternate varieties and methods could be used to increase yields, prohibitive transport costs prevented the export of low-value, high-bulk commodities. At a cost of £12.10 per ton, the price of canoe transport from the north to the coast (Tamale to Ada) was still over five times that from Ada to Liverpool. Tudhope's assessment concluded that "until a satisfactory system of transport is evolved the profitable development of almost every purely agricultural product is doomed to failure" (ADM 56/1/153).

The opposite was true of the forested southern half of the Gold Coast, where the natural distribution of valuable raw materials like gold and kola coincided with the humid conditions needed to produce the most valuable export crops: cocoa, oil palm, and rubber. The area also benefited from close proximity to the coast and more developed transport networks (Cardinall 1931, 82). Colonial agricultural policy was therefore directed at exportable commodities in the south. This resulted in so-called underdevelopment of the northern half of the country, which was effectively cut off from production of the most lucrative goods and was too far distant to make a profit from lower-value crops, given the high transport costs to the coast (Plange 1979; Sutton 1989).

Besides subsistence agriculture and craft production, the primary productive roles of people in the northern half of the country in precolonial times had been as trade intermediaries, particularly in livestock and kola nuts. Indeed, caravan tolls and market dues were the primary sources of income for the British in these areas in the beginning of their tenure there. However, the colonial administration soon sought to redirect trade through their own newly established administrative centers, such as Kintampo (below), in part by closing local markets. This, along with the coerced outmigration of men, had the effect of closing off market access to many who had formerly produced cash crops like peanuts, cotton, and rice for local sale (Plange 1979, 8–9). Though the British made some attempts to stimulate trade, such as abolition of the caravan toll in 1908, they showed little interest in the commodities actually being traded in the north after several failed experiments indicated that their international export would not be profitable. Thus, the cycle of "nondevelopment" of the north was constructed, whereby transport networks were not developed because there were no low-bulk/high-value goods, effectively preventing export of high-bulk goods since effective means of transport were lacking (Sutton 1989).

"GROW ENOUGH TO EAT AND TO SELL": BANDA IN THE EARLY COLONIAL PERIOD

Following the British-led expulsion of Samori, and after many years characterized by multiple relocations, Banda peoples resettled their former villages, including that of Makala (Kataa), a period referred to archaeologically as the Late Makala phase, from about 1897 to the late 1920s (Stahl 2001, 189). In some ways, these occupations were shells of their former selves. The village of Makala was much smaller than it had been in previous generations, at seven and a half hectares or fourteen American football fields in extent (compared to eighteen hectares in the Early Makala phase). Instead of the strong, thick-walled coursed earthen architecture of previous centuries, houses were built quickly using wattle-and-daub or pole-and-daga construction (Stahl 2001, 200). Although people had finally returned home after decades of instability, shifts in other kinds of routinized daily practices suggest that people did not inhabit spaces as before. For example, people disposed of trash differently, spreading it across the site and filling in depression-like pits rather than setting aside certain areas for accumulation (Stahl 1999b, 58–61; Stahl 2001, 200, 203). Changes like these may well have stemmed from decades of migrating, of inhabiting spaces in other peoples' territories, of living life ephemerally, as well as from the significant demographic and economic shifts associated with colonial rule.

Since Banda was situated in woodland savanna too dry for cocoa production, colonial policy and investment in the area was, as in the most of the rest of the Northern Territories, minimal. There were few British officers in the Northern Territories—just eighteen in 1905—and little was known of the Banda area. A complete village inventory, for example, was not accomplished until 1917. Attempts to mine gold or cultivate cotton as a cash crop were met with disappointment, due in part to the high cost of transporting bulk goods southwards (Stahl 2001, 194–95). The lack of direct colonial intervention might be taken to mean that colonial impacts on food and agriculture were minimal in areas like Banda, but the evidence at hand suggests otherwise. Banda's story demonstrates how important it is to look at everyday responses to colonial rule in addition to the policies of the metropole themselves. Two trends evident in Banda's material record correspond with major dietary shifts: a significant gap in labor availability, particularly of men, and increased incorporation into the market economy. In this section, I evaluate both of these shifts and their impact on foodways. A more complete rendering of daily life under colonial rule and the wider political economic context can be found in Stahl (2001, 189–214).

British officials who visited Banda in the early 1900s remarked on the low population density and generally inferior standard of living in the area. A 1907 tour of inspection by F. Fuller, the chief commissioner of Ashanti (of which Banda was now considered a part), describes the general conditions at the time:

The country is generally open and undulating; population scarce and of a lower standard from the Ashantis (if judged by their material surroundings), but more genial and docile than the Ashantis proper . . . . the tribe was unfortunately situated and suffered long from the slave raids of the Ashantis as well as from the more recent, and more thoroughly destructive expeditions of the Freebooter Samori—No wonder therefore that their numbers are comparatively few for the land they occupy and their villages small, ill-kept and generally inferior. The Pax Britannica has, however, already produced a universally beneficial effect and if the number and the condition of the children may be considered a criterion of the future condition of the race, it can be confidently asserted that the Bandas will rapidly attain prosperity.[4]

Colonial sources suggest that even in 1926 population densities were still quite low in Banda and the wider region (Stahl 2001, 198; Stahl and Anane 2011); a 1931 census puts population density in the wider region at zero to ten persons per square kilometer, among the lowest in the country (Cardinall 1931, 157). In a culture where wealth was very likely in people and the diverse skill sets they possessed (Guyer and Belinga 1995), these losses probably had a major impact on social and political life.

A lack of labor was of special concern to Banda's leaders during the early colonial period. Compounding low populations in Banda, colonial officials at the district capital of Kintampo requisitioned both food and labor from Banda, though Banda peoples were not always able or willing to fulfill those requests. For example, at the opening of the twentieth century, Banda was required to supply the district with forty-two to eighty carriers per month (Stahl 2001, 196). In 1901, the Banda chief complained that "the farms suffered in consequence of their absence, and that they themselves returned very thin and pulled down, unable to do a day's work for a considerable time afterwards" (ADM 56/1/415). While the chief may have overstated these complaints, taking men away from their farms probably did have a negative impact especially during key points in the growing season. Demands for food may also have increased over time; for example, in 1926 Banda was expected to help provision a considerable number of soldiers at Bui (Stahl 2001, 196).

People used numerous strategies, attested in the material and ethnohistoric records, to help cope with low population densities and limited labor availability. Family histories suggest that Banda families incorporated refugees and captives from different ethnolinguistic groups, which may have helped ameliorate low populations. For example, four of the seven families that founded Banda-Ahenkro, now the seat of the paramount chief, originated in ethnolinguistic groups other than the Nafana (e.g., Gonja, Gyaman). Although the chieftaincy in Banda was Nafana, in practice it comprised individuals from several different groups who adopted Nafana customs (Stahl and Anane 2011; Stahl 1991; Stahl 2001, 198–99; see also Whatley and Gillezeau 2011 on this general phenomenon).

Another strategy involved significant changes to the gendered division of labor and the crops cultivated. Both colonial demands for labor and the increasing need to access cash through labor migration would have disproportionately drawn men away from their homes and fields (Cruz 2003; Stahl and Cruz 1998, 216; Stahl 2001, 207–8). Women and elderly men potentially would have taken on a much greater share of the responsibility for making ends meet and feeding their families. In Banda, people shifted to eating maize as a staple grain (19% ubiquity), with smaller amounts of pearl millet present (12% ubiquity) and sorghum absent entirely (see appendix B). Although food preparation remains difficult to access in the material remains, charred plant remains are much sparser than in previous centuries, a change which may have resulted from an increased focus on cassava (see below) and/or from changes in food preparation.[5] It is quite possible that women abandoned grinding grains on stones in favor of the wooden mortar, which was better suited to pounding maize and cassava and remained the dominant preparation method well into the twentieth-first century. Remains of two types of wood used to make mortars have been identified in the archaeological record at the same time that grinding stones disappear (Logan and Cruz 2014, 222).[6]

These shifts in agricultural production and possibly food preparation coincide with changes in pottery production, which ethnohistorically was the domain of women (Stahl and Cruz 1998). Potters increasingly used maize cobs to roll texture onto pots, a shortcut that avoided more time-consuming decoration techniques (Logan and Cruz 2014). The diversity of decorative treatments and combinations also declined (Cruz 2003). Differences in temper in the archaeological ceramic assemblages are consistent with a pattern of reduced effort in pottery production, and perhaps increased use of wooden mortars in food and ceramic paste preparation. While all of these changes may relate to an increasing need to trade ceramics at colonial markets, they also suggest that women needed to make pots faster (Logan and Cruz 2014, 224).

Traditionally male activities diminished during the same period. Iron-working, typically a domain of men (Stahl 2016), was no longer practiced, with people instead acquiring already finished items and recycled iron from regional sources and the British metropole (Stahl 1999b, 64). Hunting, also usually a male domain, was considerably attenuated compared to Banda's earlier record, though broadly consistent with patterns in the late eighteenth to early nineteenth centuries (chapter 3; Stahl 2001, 207). Meals were less meat-focused than in previous centuries, as attested by the low density of animal bones overall (Logan and Stahl 2017, 1383). People did not eat bovids like cattle, sheep, and goat in the quantities they used to, perhaps because they were more valuable as trade goods. More difficult-to-capture animals like carnivores disappeared, suggesting the absence of skilled hunters or of the time to pursue difficult hunts (Logan and Stahl 2017, 1387), or a landscape denuded of fauna as a result of overhunting by Samori's hordes. People increasingly made up the bulk of their meat needs with animals like rodents and reptiles that could be easily captured nearby as pests in agricultural fields or villages (Logan and Stahl 2017, 1385). These kinds of changes often reflect periods of food insecurity or other stress, when people struggle to fill dietary needs with what is most accessible.

Taken together, these shifts point to women taking a larger share of responsibility in farming and in making ends meet. In Cameroon, Guyer (1978, 1980) also found an increase in female farming in association with colonial demands on men's labor. In Ghana, men were required to serve as porters and other posts by colonial governments, and were encouraged to migrate southwards to earn cash as laborers on cocoa farms and in gold mines (Cruz 2003; Grier 1981; Hill 1963; Stahl and Cruz 1998). This emphasis simultaneously removed men from family labor obligations and firmly installed the infrastructure of the market.

Lower availability of labor must have taken a toll on the amount of land people were able to farm, and it reduced the quantities of food people were able to produce. In this low-labor context, cassava makes the most sense as a replacement crop, because it produces exponentially more calories per unit labor than any grain crop (Jones 1959). While cassava is hard to recover archaeologically, archival sources reveal that it was present in the broader Volta basin by 1894–95 (Stahl 2001, 208). We also know that cassava was widespread in Kintampo, the market town visited by Banda women, by 1902, where it was considered "poor, coarse food" (ADM 56/1/458). Both of these earlier indications, in addition to the changes in food preparation mentioned above, suggest that cassava was also in Banda at the turn of the century. Archival sources reveal that cassava was widespread in Banda by 1931 (BRG 28/2/5). Although we cannot definitively say when cassava became a staple in Banda, it seems unlikely to have been chosen out of culinary preference. Instead, Banda's adoption of cassava indicates the consumption of a less-preferred food associated with poverty, and provides another line of evidence for the food security challenges people experienced during the first few decades of the twentieth century. Cassava continues to provide cheap and accessible calories in Banda today, where its consumption has increased markedly in recent decades (chapter 5).

Although maize may have reduced the time spent processing and preparing food, compared to cassava it was quite expensive to produce if measured in calories produced per unit labor (Miracle 1966). Thinking more broadly, there are several additional factors that might explain maize's increased importance in Banda, including changing tastes as well as the desire for a fast-growing crop that matures during the hungry season. Maize might have been considered more desirable than in previous centuries, although there are no clear indications that this is the case. The co-occurrence of cassava as well as the consumption of low-ranked animals suggests that at least some people in Banda may have not been able to consume preferred foods. The distribution of maize across archaeological contexts is not restricted as in early periods, so it seems unlikely that maize was a food enjoyed only by elites or other special groups. Maize's use as a staple probably has much more to do with its ability to produce a harvest quickly, before yams and grain crops are ready to harvest. Only fast-maturing pearl millet can produce a crop in as little time as maize, but millet requires a much greater expenditure of labor for cultivation (Miracle 1966) and processing. What is perhaps most significant about maize's quick maturity is that it becomes ready to harvest during the worst part of the hungry season, when food supplies from the previous year's grain harvest are often depleted and just before yams are ready to harvest. For these reasons, maize is in many ways a barometer of the hungry season, and its proliferation in late nineteenth- and early twentieth-century Banda suggests that this particular kind of scarcity may have become more pronounced at this time.

Why might the hungry season have become more marked at this time? I suspect it has much to do with the increasing entrenchment of market economies, a complex dynamic that requires some unraveling. Food was required to supply growing populations of nonfarmers in colonial centers (Guyer 1978). Banda, for example, was expected to supply food to Kintampo, the district headquarters (ADM 56/1/458; Stahl 2001, 196). Selling agricultural produce was also one of the few ways in which rural villagers could access cash. Yet these demands on food supplies came at the same time that labor was in short supply, and agricultural production was likely at its lowest, as Banda peoples resettled in the area after decades of dislocation. Even if families could have stretched already diminished food stocks through the year, the need to sell a portion of their harvest would have drastically reduced their ability to feed themselves. This tension may have been why rural people were reticent to supply Kintampo's market (Stahl 2001, 196).

These dynamics may have resulted in the entrenchment of the hungry season gap, which is often assumed to be a natural feature of African agricultural cycles. The hungry season gap occurs during the growing season, when last year's harvest runs short and the current crop is not yet ready to harvest. Prices for food are high at this time, since demand for food is also high, a dynamic which is observed in monthly records of food prices in turn-of-the-century Kintampo (ADM 56/1/415). In May 1901, the beginning of the growing season, a colonial officer notes that there "is very little food" in the district. By July 1901 the prices of food had gone down a little, with only "Kassava [sic], young corn, young yam, and spinach" being available at reasonable prices. Note here that it was American crops, maize and cassava, that were available, as well as young yams that had probably not reached their full size. By August/September, prices were half what they had been just months earlier, and there was a much wider diversity of foods available, including plentiful yams. This description of food availability and pricing follows the growing season of African crops, and indicates a lack of food in the early growing season known as the hungry season gap. While people with means, such as the British authors of the report, could have afforded to pay twice as much for food, villagers would have been especially hard pressed to make ends meet.

Women may have been active in mobilizing their social connections to help deal with seasonal food shortages. One place we can trace these networks is in the pottery women produced and consumed. The movement of pots, from producer to consumer, signals economic and social relationships. Sourcing studies of pottery found in Banda at this time indicate a greater variety of production centers from both east and west of the hills. The most abrupt change is a shift away from acquiring vessels made on the west side of the hills and increasing consumption of vessels from closer contacts on the east side. These patterns may relate to continued risks and uncertainty in traveling to and from the west, or to the influx of new refugee potters (Stahl 2001, 205–6; Stahl et al. 2008). In any case, they also signal exchange among a closer set of women and men, with whom face-to-face interactions were more common. Food may have been a central part of these relationships. Ethnohistorically, pots were often bartered for their volumetric equivalent in foodstuffs. Producing more pots may have provided a means to access food supplies in lean times. Building these relationships with smaller, more intimate networks would have been an excellent social strategy, since people are much more likely to share food with people they know and can call upon in later times of need.

The British were actively engaged in creating markets for their manufactured goods, a process that was seen as not only benefiting the Crown's coffers but also acting as a stimulant to improve agricultural production in the colonies. Cash was required to purchase such goods, and for most rural people outside major cash-cropping zones in the south, growing enough agricultural surplus was the primary means to access it. W.S.D. Tudhope (ADM 56/1/153), on a 1912 tour of the Northern Territories, including areas immediately across the river from Banda, described these hopes:

> A desire for money appears to have been awakened and they are beginning to realise the additional luxuries that money can provide for them. This indicates probably a surer sign than any of the possibility of a further development in Agricultural productions,—the only source of wealth open to them in their own country. Many of their young men against the wishes of their relatives have left the country to hire themselves as labourers which they would not do if they could get remunerative employment at home . . . A considerable proportion of the male population are employed on transport and other Government work . . .

Yet while hoping the desire for consumer goods would stimulate production, Tudhope himself also acknowledged the futility of this endeavor due to high transport costs (above). The potential for cash cropping was explored in all corners of the Gold Coast, and most studies of the Northern Territories came back with assessments similar to Tudhope's.[7] Numerous sources attest to the growing of cotton in the Banda area beginning in 1904 (ADM 56/1/421), though the origin of its cultivation was probably much older. In 1928, colonial officials evaluated Banda and areas west for the production of cotton as a cash crop, but were sorely disappointed regarding yields, since cotton was intercropped. Farmers were resistant to growing monocrops of cotton, and the British soon abandoned the idea. Instead, they favored the production of groundnuts (peanuts), which were produced in enough quantity to be sold at market (BRG 28/2/5). Other than encouragement of these two crops, as well as urging Banda farmers to supply the Kintampo market with foodstuffs, there seems to have been little direct colonial interventions in agriculture in Banda before 1930.[8]

Yet the indirect impacts of market policies and monetization were pervasive. Selling off agricultural produce to local markets was one of the few ways in which rural people could access much-needed cash. This meant that people had to "grow enough to eat and sell," a phrase used today to describe the most common strategy for accessing cash (chapter 5). Until recently, farmers have focused on growing subsistence crops, hoping their harvests will be of sufficient quantity to not only feed their families but also sell at market. But harvests are rarely sufficient for this purpose, at least on any sustainable scale, and certainly production on this level would have been difficult if not impossible given low labor conditions in the early colonial period. A surplus harvest one year may be followed by a deficit the next, and the margin of surplus is often so tiny that it does not account for the need to purchase foodstuffs at high prices during the hungry season.

Cash was needed not only to purchase items at markets but also, increasingly, to pay taxes and other colonial obligations. The British actively engaged in creating local markets for their manufactured goods, initially stocking government stores with these items to encourage desire for them (Stahl 2001, 195). In the first decades of colonial rule, the largest and closest market town and colonial center was Kintampo, just under eighty kilometers away, about a two- or three-day walk; later Wenchi, which is slightly closer, became the district center. People from Banda would have had to go to Kintampo to conduct official business like court cases, and to reach a wider market for their wares. Archival sources suggest that Banda women sold their goods there on an occasional basis (ADM 56/1/415).

What was available in Kintampo, and what of these items did Banda peoples choose to purchase? A brief survey allows us to indirectly assess the degree to which Banda's inhabitants were integrated into the British colonial market economy. In 1901, Kintampo's market offered cloth, basins, pans, knives, fezzes, wooden pipes, towels, pomade, snake beads, looking glasses, combs, matches, sugar, brass and copper rods, lead bars, and soap (ADM 56/1/415). Several of these items and more are found in Banda's refuse piles, attesting to the replacement of some local industries with imported alternatives. European-made ball clay pipes were among the first imports to be used regularly, and they soon replaced local alternatives (Stahl 2001, 206). Pipes may have also been obtained through barter—perhaps of tobacco, which we know Banda peoples exchanged at the Kintampo market (below) (Stahl 1999b, 63, 70).

Banda peoples enjoyed other imported substances like bottled spirits in some quantity. Medicines and pomades were also consumed, as attested by a smaller quantity of milky, clear, and blue glass. These objects stress the global connectedness of places like Banda, probably mediated through the market at Kintampo (Stahl 1999b; Stahl 2001, 195–96, 209–11). These items suggest that Banda villagers were able to obtain petty cash, probably through trade of local produce such as tobacco, groundnuts, or cotton. They also suggest elaboration on existing tastes for body oils (Vaseline, for example, may have been a stand-in for local shea butter), for new medicines to cure ailments of the elderly, and for alcohol, the consumption of which was taken up with gusto for reifying the authority of chiefs (Akyeampong 1996; Stahl 2001, 209–11; Stahl 2002).

Imports may also have provided convenient replacements for labor and skills that were diminished after Banda's demographic collapse. Manufactured ointments would have saved women a lot of work, as shea nuts can be time-consuming to collect if the stands are far away, and the oil is laborious to extract and process. Imported alcohol may have also released women from the need to produce large quantities of beer. The presence of imported metals like iron and brass (Stahl 1999b, 64,70) might have helped make up for the lack of men with either the knowledge or time to produce these items at home. Other items, like gunflints, present even earlier than the colonial period, attest to the introduction of new technologies that spread rapidly and had major consequences for hunting and warfare (Stahl 1999b, 70).

Still other industries proved resilient against imported alternatives, particularly ceramics and cloth. Although imported vessels were readily available, as attested by the very small amounts of imported white ware and a metal enamel vessel present in the material remains (Stahl 1999b, 64), Banda specialists continued to practice potting and most people used these local pots (Stahl 2001, 207; Stahl et al. 2008), as discussed above. People also continued to produce cloth locally, instead of switching to the manufactured alternatives that were rapidly making inroads. Although Banda peoples controlled the entire supply chain for cloth, from farming cotton, to spinning thread, weaving, and dying, the scale of production appears to have been limited to household-level production (Stahl and Cruz 1998). One of the main challenges in estimating scale is that spindle whorls are often curated and passed down from mother to daughter, thus they rarely appear in the archaeological record (only nine are present in Late Makala contexts; Stahl 2001, 204). One seed that is likely indigo was recovered, not far from a feature that resembles a dye pit (Stahl 2001, 203).

Yet other imported goods simply replaced items formerly imported from elsewhere, though not without consequences for how they were used. Beads are a case in point, since they had long been acquired from some distance away (chapter 2; Stahl 2001). In Late Makala times the source of these beads shifted, with about half of those recovered coming from Europe. Some of the more exquisite multicolored hand-drawn and wire-wound forms, which resembled earlier forms, were used in sacred contexts like nubility rites. Cheaper, mass-produced monochrome forms were also adopted and used for more mundane purposes (Stahl 2001, 211).

The market also provided opportunities to access other kinds of food from a wider world, and attention to it brings into focus the foodscape of central Ghana, even if Banda diets were probably considerably more provincial. In 1902, British doctor W. Graham meticulously documented the foods available at Kintampo (ADM 56/1/458). The Hausa, an ethnic group originating in Nigeria and apparently working for the British at Kintampo, preferred maize in the form of *kenkey*, a fermented dough wrapped in leaves. Other African groups did not like the sour taste of *kenkey* and instead preferred *fufu*, preferably made of yams but sometimes plantain, cocoyam, and/or cassava. Even today, while people in Banda have now developed a taste for *kenkey*, most women do not make it regularly at home.

In Banda, *fufu* made of yams was probably preferred as it was in Kintampo and continues to be today, but in practice this dish was probably also made out of cassava. *Fufu* is a seasonal dish whose consumption is linked to the availability of yams, starting in August until they are finished. Cassava, however, can be obtained year-round, and would have extended the seasonal availability of this dish. *Tuo zafi* made of maize and pearl millet likely formed the dietary mainstay of people's diets in the dry season and early wet season (November through July). Cassava can also be ground into a flour and used to make *tuo zafi*. Graham does not mention this northern dish at Kintampo, but its preparation and widespread consumption is well attested in Fortes and Fortes's (1936; see below) ethnography of the Tallensi in the 1930s as well as in oral accounts from Banda (chapter 5).

Each of these starchy staples is served today with what in Ghanaian English is called "soup," a phrasing and culinary grammar that was also common in turn-of-the-century Kintampo. This "soup" is more akin to a thin sauce in American English usage. People particularly enjoyed peanuts, prepared into a soup or simply roasted. While peanuts have not been attested in Banda's archaeological record, Russell (BRG 28/2/5) notes that they were cultivated as a cash crop there, perhaps for trade at Kintampo; it is unclear whether people commonly consumed peanut soup. Other soup ingredients were common at Kintampo, including the leaves of the sweet potato plant, tomatoes, okra, and eggplants. Soups were flavored by local ingredients like *gabu*, pulverized onion leaves, and *dawadawa* balls, both recalled with fondness in oral food histories in Banda and likely available in the early twentieth century (chapter 5). Other spices sold at Kintampo included a local variety of ginger, chili peppers, black pepper, and *melegueta* pepper. The range of soup ingredients found in Banda's archaeological record is much smaller, though this is probably a function of preservation rather than a lack of availability or preference. Okra and several wild leafy greens (*Cassia* spp., *Portulaca* spp., *Zaleya pentandra*, and cheno-ams) were found in trash deposits, suggesting the continued preference for a slippery texture as well as the taste and nutritional affordances of wild greens.

Many different kinds of beans were available, chief among them cowpea, also identified archaeologically at Banda, and Bambara bean. In Kintampo and probably in Banda, most women used shea butter for cooking, though palm oil was available for those with the taste and means to access it. A wide array of tropical fruits were also available, including pineapples, limes, oranges, bananas, mangoes, and papayas, several of which can also be grown in the Banda area. Tobacco was among the market's popular items (ADM 56/1/458), and it was certainly consumed by Banda peoples judging from the prevalence of tobacco pipes.

Sorghum was also widely cultivated near Kintampo and prepared into porridge or *pito*, a local beer (ADM 56/1/458). Beer brewed of an unspecified grain was made locally according to reports of the 1906 trial of Adjua of Banda, who was charged with poisoning her brew to murder another woman (ADM 56/1/421). *Pito* is today made out of sorghum, but sorghum disappears from the plant repertoire in Late Makala deposits. This may simply be a preservation issue, or beer may have instead been made from pearl millet or maize, as is occasionally done elsewhere. Stahl (1999b, 69) suggests that the intrusion of foreign beer and spirits may have eroded local beer production, though Adjua's story suggests that it was still common enough, and indeed it continues today.

It is difficult to evaluate how these shifts in food availability and access might have influenced people's health and well-being. While such information can be gained from human remains, per an agreement with local Banda communities we do not exhume ancestors during the course of archaeological investigations. However, height data suggest that food deficiencies occurred elsewhere in the colony during the early colonial period in particular. Height is not a perfect indicator of health and nutrition; it relates most strongly to protein consumption. Austin and colleagues (2007) found a significant decline in heights in people born in Ghana between 1895 and 1909. While correlation is not causation, it is worth noting that the early part of this time frame encompassed three wars between the British and Asante, as well as the British taking formal control of the new Gold Coast colony in 1901. Violence often leads to severe food insecurity, as the Banda data also suggest. But these effects persist well into the first decade of the twentieth century, long after the cessation of direct violence. Note that this period corresponds precisely to when we see Banda's inhabitants switch to maize as a staple, in the period encompassing the 1890s to 1920s. One of maize's primary roles in modern West Africa is in plugging what in the twentieth century became known as the hungry season gap, given that it matures precisely when people run short of last year's harvest. A significant decline in height, as proxy for health, as well as the switch to maize as a staple in Banda in the 1890s–1920s—just as colonial officials were forcing engagement with cash—may well signal that the hungry season gap had become a more or less permanent feature of agricultural cycles in affected areas by this time.

SCIENCE AND THE "DISCOVERY" OF MALNUTRITION IN THE LATE COLONIAL PERIOD

For many decades, the challenges of feeding one's family were simply beyond the gaze of the colonial government, but this began to change in the interwar period (Worboys 1988). This policy shift was coincident with growing faith in the use of science to understand and improve the welfare of the colonial citizenry (Hodge 2011; Tilley 2011). In British Africa, these developments were closely linked to changing perspectives on the role of the colonial state as well as African labor.

Labor had long been a limiting factor in the continent, and it was thought that improving the health of Africans might help increase the quality and size of the labor force. How these changes in policies played out on the ground varied (Iliffe 1987; Tilley 2011). In many colonial contexts, malnutrition was framed as a scientific problem rather than a symptom of the political economy of colonialism, a view which continues to resonate in many development circles and has the effect of depoliticizing food distribution. Other colonial settings, the Gold Coast among them, mostly ignored the growing evidence of malnutrition in their colonies and did not enact comprehensive nutritional policies (Robins 2018, 168).

In the Gold Coast, Africans residing on the coast were more accessible to the colonial gaze, and some of them consequently came under closer scrutiny. By contrast with the north, there was abundant food in the south, but food costs there were high. For urbanites with steady employment, wage increases made up for increased food prices. At the same time, food imports, particularly of rice, flour, and tinned meat, fish, and milk, were on the increase (Robins 2018, 170). Literate urban Africans enjoyed these high-status foods and consumed them in abundance. The colonial government should have been pleased by this development, since it meant they earned more on imported food tariffs. But instead, beginning in the 1920s, they attacked African consumption of imported foods as a lazy, unhealthy habit (Robins 2018, 171). One can only speculate as to why colonial officials critiqued Africans for eating the same foods that the British did, but I suspect it has a lot to do with policing of racial boundaries. In response, numerous attempts were made to increase the consumption of local foods among African elites. This was done under the guise of improving health, since imported foods were deemed less nutritious (Field 1931; Robins 2018, 172–73). Still, agricultural policy in the 1920s Gold Coast remained focused on cash crop production (Robins 2018, 174).

In 1926, William Ormsby-Gore, undersecretary of state to the colonies, toured West Africa and vocalized concerns that cash cropping had been too successful and had reduced the arable land devoted to food crops. He was concerned with the nutritional quality of foods, and like his counterparts in the Gold Coast expressed growing concern over Africans' consumption of imported foods. To him, a nutritious diet was critical for nurturing a productive labor force. He promoted the idea that food should come first, and economic concerns second, though his connection of diet to labor reveals that economic concerns remained paramount in his thought process. Growing concern with African welfare among reformist groups in Britain in the 1930s led to the establishment of the Committee on Nutrition in the Colonial Empire, which published a 1939 report declaring that there was no major food shortage in the British Empire. Instead, Africans' poor nutrition was blamed on African laziness and the lack of variety in their foodways—in short, a poor diet was their choice (Robins 2018, 175).

Given the time and effort expended to alter the diet of the urban African elite, who did not suffer from food insecurity, one might expect at least similar attention to the diets of the poorest Africans. This did not happen until the 1930s, when new scientific methods in the form of the nutritional survey led to the discovery of malnutrition in the colonies (Worboys 1988). In the Gold Coast, medical officer F.M. Purcell was tasked with making a nutritional survey of the entire Gold Coast. His grisly 1939 illustrated catalog of deficiency diseases near Akim, in forest country, connected these ailments to nutrition, and showed the impact of high food prices on the rural poor. In subsequent drafts of his national survey he found widespread nutritional deficiencies in the Northern Territories, where people were lethargic due to low food intake during times of "mild famine" (the hungry season) and suffered frequently from scurvy. He characterized the Gold Coast as a whole as "poor in indigenous foodstuffs." Noting the importance of introduced foods like maize and cassava, he suggested the introduction of more foreign crops. To further improve nutrition, he thought women ought to be taught how to properly feed and cook for their families. He viewed pounded starchy staples like *fufu* with particular derision, noting that they involved very arduous work for women, and that their manner of preparation was nothing short of "violent," inducing of course equally untoward health effects (Robins 2018, 175–77).

Unlike his Akim work, Purcell's national nutritional survey, which reported severe malnutrition in the north, was never published; indeed, it was actively suppressed. A British official reportedly told him "no one may starve in the British Empire," revealing the colonial government's concern that the report would fall into the hands of the empire's critics. Purcell resigned in protest in 1943, but his findings and recommendations do seem to have had an influence on later developments (Robins 2018, 177). By the 1940s and 1950s, the British colonial government promoted a Grow More Food campaign in the Gold Coast, and made food production a priority over cash crop production. Education campaigns for women were less effective, not because women did not readily take to their lessons, but because they preferred the taste of traditionally prepared foodstuffs. Many of these policy changes and recommendations were continued in the early years of independent Ghana (Robins 2018, 178–79).

During the same years, the science of anthropology was also employed by colonial governments as a means to understand their subjects (see Tilley 2011). The ethnographic work of Meyer Fortes and Sonia Fortes among the Tallensi in northern Ghana, published in 1936, helped to flesh out the local context of food shortages, which Purcell was unable to gather using nutritional surveys alone. Unfortunately, we do not have archaeological data available for the late colonial period in Banda, since those deposits are buried under modern occupation. The Tallensi study brings to life the experience of the hungry season for this far northern region, with implications for understanding life in Banda during these decades. I agree with Shaffer (2017) that these experiences should be evaluated in a highly localized fashion, so my description of the Tallensi should not be understood as an interpretation of Banda's specific circumstances, even though Banda was similarly situated in relation to the colonial economy and there are considerable similarities to recollections of "olden times Banda" (chapter 5). In particular, the drier environment occupied by the Tallensi would have afforded a much smaller margin of security than in Banda.

Among the Tallensi, the primary foods were pearl millet (early- and late-maturing varieties), sorghum, and sometimes rice. Maize was not grown locally but was consumed occasionally when it was available cheaply at market. The emphasis was on crops that matured at different times of year so as to insure the food supply year-round. Groundnuts were frequently cultivated and often sold, and appear to have been the only significant cash crop. Other crops included Bambara beans, cowpeas, Kersting's groundnut (*Kerstingiella geocarpa*), *frafra* potatoes (*Coleus dystenericus*), and sweet potatoes. Vegetables were the domain of women, and included *naŋgena* (*Gyandropsis pentaphylla*), okra, *bɛt* (*Hibiscus sabdariffa*), *bɛris* (*H. cannibus*), and *neri* seeds (*Cucumis melo*) (Fortes and Fortes 1936, 242–45). Meat consumption was limited and unpredictable, even if almost all households possessed fowls, most had sheep and or goats, and a few of the more wealthy households kept a cow or two. The animals were slaughtered rarely, particularly in the case of cattle, which served as bride wealth. Animals were viewed as mobile wealth rather than as providing a steady supply of meat, eggs, or milk.[10] Shea butter was the only oil. Flavorings included *dawadawa* balls, red pepper (often imported), rock salt from the south, as well as ashes prepared from grasses and used as a salt substitute (Fortes and Fortes 1936, 248–51).

Fortes and Fortes (1936) portray Tallensi life as vulnerable and highly susceptible to famine. In their original article, they capture the food cycle in a two-column format that highlights the seasonality of food availability and some of the social strategies people used to access food in trying times. I reproduce this information in full detail in table 1, because it captures the dynamics of food insecurity, which are best understood in a cyclical way (see also Mandala 2005).

Several elements of the Tallensi case study are pertinent to understanding the dynamics of food insecurity in the 1930s and how people managed to avoid it. People employed a number of social and economic strategies to make ends meet. These strategies not only tell us how people survived, but also provide clues into the severity of hunger. We know from modern studies that people suffering from food insecurity employ a hierarchy of coping strategies (Devereux 2001; De Waal 1989). People preferentially use strategies that have only short-term impacts, and avoid doing things that will impact their long-term survival unless they are out of other options. In other words, people often choose to go hungry in the short-term to preserve their long-term assets. Livestock are one of the most widespread long-term assets; they are usually not sold or slaughtered unless the owners are suffering from severe hunger. That Fortes and Fortes (1936) mention this method suggests at least some Tallensi were in a particularly bad state.

Given a choice, most Tallensi employed less costly economic and social strategies to mitigate food shortages rather than consume or sell off their long-term assets. If money or barter goods were available, people could trade them for grain

TABLE 1 Tallensi Productive and Food Cycles in 1934 (abridged from Fortes and Fortes 1936, 253–59)

APRIL: First planting-rain fell on March 26, and early millet with some cow-peas and *neri* was planted in valley settlements. Rain again in mid-April enabled most people to plant early millet, &c. Women sowed their vegetable patches. . . . Hoeing begun; but not more than one day in three devoted to agriculture. Communal fishing expeditions.	Food stores very low in average households and being rationed. Many dependent upon supplementary sources of supply. People buying grain abroad [from other regions] for re-sale. Ample food supplies in market and many buying grain. Market-prices average. Children gleaning ground-nuts. Wild fruits . . . being consumed to stave off hunger.
MAY: Rains continue erratically. Early millet planted in the stony and hilly areas. Others interplanting guinea-corn and some ground-nuts. Work on bush farms preparatory to planting commenced. Tempo of agricultural activity increasing rapidly.	Food stores deplenished and severe rationing. "Hunger" (*kɔm*) commences. Poorer households suffer two or three days' hunger a week, living on vegetable soup, groundnuts and wild fruits. Householders send their wives to purchase grain abroad [from other regions] for consumption and re-sale if they have money. Many selling live stock bit by bit to buy grain in market. Prices of all commodities rising. Visits being paid to relations in more fortunate areas to get some grain.
JUNE: Height of agricultural season. Men completely absorbed in hoeing and weeding, planting guinea-corn and late millet on bush farms, rice, ground-nuts and minor crops. Hiring labour for help in hoeing and weeding . . .	Peak of hunger reached. Granaries empty among poorer households. Much live stock sold or bartered for grain very cheaply; grain scarce and dear. Much ground-nuts for sale. Small groups of children wander about hungry, feeding on wild fruits and small animals they find near the settlements. Towards the end of the month many people are staunching their hunger by cutting the ripe or half ripe heads of early millet which they roast on the embers and nibble at.
JULY: Early millet harvested by those who planted first on valley land. Hoeing and weeding of compound and bush farms in full swing. Some still planting subsidiary crops. Poultry breeding begins.	Early millet eaten by those who harvest. Relatives from the later-planting areas come to beg grain. Wild fruits still eaten and hunger prevalent in late-planting areas. Ground-nuts, beginning to ripen, plucked and chewed as snacks.
AUGUST: Late planters harvest early millet. Invited collective labour for hoeing and weeding coming to fore, but hired labour still in evidence. Fresh hibiscus leaves used in soup.	Almost everybody has early millet; hunger appeased for a time.
SEPTEMBER: Harvesting ground-nuts and other subsidiary crops (roots and legumes) begun. Hoeing and weeding of bush farms continue. Women plucking green *ocro* and burning early millet stalks for *bakaa* [salt substitute]. Ritual festivals commence.	Those who harvested in July already reaching end of supplies of early millet, and resort to market and relatives. Early millet very expensive in market. New ground-nuts (partly green), roots, and legumes supplementing diet largely.

TABLE 1 *(Continued)*

Productive Cycle	Food Cycle
OCTOBER: Harvesting of guinea-corn, rice, root crops, legumes, &c., general. *Neri* lifted. Women drying and storing vegetables.	Food becoming plentiful. Price of grain falling rapidly in market. Cooked food cheap and plentiful in market. Marriage season begins.
NOVEMBER: Harvesting late millet on bush farms, cow-peas and Bambara beans. Women burning valley grass for *ziem*. Children gleaning ground-nuts in play, not because of hunger. Height of ritual festival season.	All kinds of food plentiful. Harvest festivals, during which enormous quantities of food are consumed and circulated and many animals sacrificed, giving the maximum meat supply of the year, and root crops including purchased yams especially in demand for the festivals. All foodstuffs abundant and cheap at market.
DECEMBER: Agricultural work over.	Height of ritual season: funeral ceremonies, children's dedication ceremonies, many marriages; all involving consumption of live stock and grain, in sacrifices, as cooked food, and beer.
JANUARY: Period of secondary activities such as housebuilding, cutting grass and timber for roofing, handicraft work. Young men begin to go abroad temporarily.	Grain rationing again begins. Grain and other foodstuffs, including imported maize, plentiful and cheap at market.
FEBRUARY: As in January.	Grain issued for only one meal a day. Many purchasing grain rather than using up their own supplies. Ground-nuts and cow-peas regular and important supplements to cereal food. Women finding grain to augment supplies, by trading, etc. Market supplies still plentiful, prices average.
MARCH: As in February. Sporadic cleaning of fields and hoeing.	Grain supplies sinking and more carefully rationed. Children and women gleaning ground-nuts to stop hunger. People going to Mampuru and Nabte countries to purchase grain for re-sale. Grain still plentiful in market, also other products, e.g. yams (imported).

at the market. Most often, this included selling value-added prepared foods like *dawadawa* powder, collected and cultivated vegetables, as well as peanuts, which were the primary cash crop. Livestock were bartered only if there was nothing left in the common stores. Grain was also rationed frugally to stretch out meager supplies, beginning often quite early in the season. Women with some means also often traveled moderate distances to barter for grain, which was then sold locally in both raw and prepared forms. Wild foods, like shea butter tree fruits, *dawadawa*, baobab, and *Celtis integrifolia*, were also important supplements (Fortes and Fortes 1936, 246–47). Importantly, many of these tasks involve the conversion of labor, especially that of women, into cash.

The food cycle relayed by Fortes and Fortes (1936) is depicted side by side with the productive or agricultural cycle, as the two are closely linked, but it would be a mistake to assume, as many scholars have, that the agricultural cycle is sufficient to explain the hungry season gap. Nowhere do Fortes and Fortes mention an absolute lack of food supplies in the wider region. Instead, these data also tell of another correspondence: between the price of food and the hungry season. Prices for food were higher, even double, during the hungry season months. High prices put food out of reach for the poorest people, who, as discussed above, used increasingly desperate strategies to access the cash needed to acquire food. Once people were forced to liquidate their long-term assets, they become embroiled in recurrent cycles of hunger.

COLONIAL INTERVENTIONS AND THE ENTRENCHMENT OF THE HUNGRY SEASON GAP

As one of the few places on the continent with data on precolonial food security, Banda provides us with a unique opportunity to interrogate the impact of colonial rule. From what we can tell, food security was significantly higher prior to European interventions. As discussed in chapter 2, Banda maintained a high level of food security during the drought of the fifteenth through seventeenth centuries, setting the bar much higher than the late nineteenth-century referent often used as a precolonial baseline. In the late eighteenth century and the first decades of the nineteenth century, although Asante demands redirected trade and siphoned at least some wealth from Banda, food privations seem to have been experienced only sporadically, during military engagements or upon capture. What is perhaps most telling in both of these cases is the minimal role played by maize, which suggests that its affordances, like the readiness of harvest before yams, were not central to meeting peoples' food security needs. Its relatively low usage suggests that such chronic shortages were not as commonplace as they became in the early colonial period. This argument has major implications for understanding why maize was adopted at variable rates across the continent. Rather than assuming that Africans were in need of maize's caloric and agronomic affordances, I have demonstrated that those needs were already met by indigenous grains, and that choice rather than privation defined precolonial Africans' choice of what to eat and grow. For most of the centuries in which it has been available, maize has been a solution in search of a problem.

In this chapter, I argue that the central problem—the hungry season gap—seems to have become more severe and entrenched in the early colonial period. The early colonial period was certainly not the first time that people experienced hunger in Banda, Ghana, or wider West Africa. Both Shaffer (2017) and La Fleur (2012, 137–44) trace occurrences of famine in precolonial historical records, usually as a result of a severe environmental shock or war and violence, the latter relayed in oral histories in Banda as well. As Iliffe (1987) argues, famines seem to have been far more common throughout the premodern world—not just in

Africa—prior to the development of steam and motorized transport, which allow states to move large quantities of food to areas most in need. But what about the seasonal, chronic hunger described in this chapter, which continues to affect far more people far more often than these severe events? Unfortunately the visibility of the hungry season in the records is much lower than that of severe events like famine, precisely because people have long managed to cope with predictable, chronic shortfalls in food supply and their effects are less noticeable than those of major food shortages. This relative invisibility led Watts (2013) to refer to seasonal, chronic hunger as "silent violence." If seasonal hunger flew beneath the colonial radar, we can also imagine that earlier chroniclers might not have recognized the signs, which might explain its absence in part in early documentary sources.

But it is possible to trace the hungry season in precolonial times and more work needs to be done to evaluate its severity and geographic spread. Oral historical and ethnographic accounts provide one major archive, and much more work could be done using historical linguistics to trace concepts of hunger back in time. Price records also suggest highs and lows in demand for food crops and reflect shortages; more might be done with earlier records from ship captains along the coast. As I've argued above, tracing the prevalence of maize and perhaps cassava provides another possibility, particularly in areas where they are known but not adopted with gusto. In these cases, it is critical that we also trace the specific historical contexts of maize adoption. Where its proliferation coincides with other warning signs, such as economic and demographic slumps, as in Banda, a stronger argument might be made for the existence of seasonal hunger.

The Banda case argues for a late emergence of chronic, seasonal hunger, which seems to become more pronounced in the early colonial period. But does this mean colonial policies are to blame for chronic hunger? And if so, how? This is a difficult problem to unravel because the dynamics at play are complex. The limitation of archive in this case means that sources that derive from the early colonial period are often used to argue for an earlier presence of the hunger season, as in Shaffer's (2017) argument. This issue is exacerbated by the structural functionalist leanings of the anthropologists who provided some of the most detailed and earliest records of rural agriculturalists like the Tallensi and the Bemba. Moore and Vaughan (1994) argue that Audrey Richards collapsed all changes prior to her arrival among the Bemba in the 1930s, even though the region's agricultural base had long been impacted by and responded to shifts in British policy. Fortes and Fortes adopt a similar perspective in some of their writings on the Tallensi, despite the fact that that group had been under British authority for forty years by the time they were made anthropological subjects. And British rule was hardly the first political change of guard that impacted foodways, as this book demonstrates. The dynamics that underlay the extension of formal colonial rule have their roots far earlier in time, and also took place in a volatile landscape that was anything but representative of the precolonial past.

To understand the impact of colonial rule on African rural economies, most scholars have attempted to evaluate whether there is a direct link between specific colonial policies and African agricultural production or food security. This tactic would ideally allow for a clear cause-and-effect argument. Fortes and Shaffer, for example, evaluate the role of labor migration to the south, and conclude it is not of sufficient scale to explain Tallensi hunger. Unfortunately, this rather strict test of colonial impact discounts how multiple types of colonially imposed strategies interacted on the ground, a situation in which the colonial government's laissez faire attitude let changes run amok. Colonial economic and policy shifts did not happen in a vacuum, but against a backdrop of complex histories and local politics. While disentangling cause and effect becomes murkier in these situations, it is important to take this broader view, since colonial interventions likely had indirect effects on far greater numbers of people.

Consider, for example, Shaffer's (2017, 285) contention that changes in the social safety net or moral economy were factors "unrelated" to colonial polices, therefore outside of his attempt to evaluate colonialism's impact on food security. He is correct in that there was no official colonial policy that forbade people from sharing food with one another. Yet historian Elias Mandala and geographer Michael Watts have illustrated the strong impact of colonial economic policies on these social safety nets in Malawi and Nigeria, respectively. The emphasis on cash crop production meant that people devoted some resources formerly reserved for food production to new and often nonessential crops. This dynamic was especially alarming to colonial officers like Ormsby-Gore, who were concerned with African nutrition. While cash cropping was less important in Banda and Taleland than in cocoa-growing areas, both areas did rely on the cultivation and sale of smaller-scale cash crops like peanuts (Fortes and Fortes 1936). These seemingly subtle differences had a variety of implications for food security.

In Banda's case, there are multiple possible explanations for why food insecurity, and particularly the hunger season, became more pronounced in the late nineteenth and early twentieth centuries. While I make the case that some of these arguments better account for the available data, I should emphasize that all of these dynamics were probably at play at various points during the early colonial period. One possibility concerns a decline in the availability of farm labor. Colonial demands for labor, as well as labor migration, occurred in a context of very low population density. Yet it seems probable that Banda farmers adopted cassava to mitigate these challenges, and in this way may have staved off the worst shortages. Low labor availability also does not account for the increased presence of maize, which requires about the same amount of labor as does sorghum, though less than pearl millet.

While a paucity of labor was likely a contributing factor to Banda's food situation, market integration poses a much more serious problem. In Banda we have multiple material indicators of market activity, including British goods like pipes, pomades, and coinage. Banda residents must have traded something to acquire these goods. Archival sources point to Banda women selling food at Kintampo's market. We also have the recorded complaints lodged that Banda was not able to supply the required quantity of food produce to the colonial government at Kintampo. The selling of food crops to access cash is a significant economic shift with direct implications for food security. Even if Banda farmers were able to produce the same quantities of food that they had in the past (unlikely, given labor concerns), selling some at market reduced the amount available for consumption. The immediate consequence is that food supplies that would have normally gotten their families through the year likely ran short before the next year's harvest was ready. When food supplies ran short, the market was an option for people, but as colonial records indicate, prices of food were often double the norm during periods of high demand, particularly the hungry season. Banda's residents were mostly subsistence farmers and small-scale traders, so the lack of a lucrative cash crop like cocoa meant low profit margins overall. No wonder maize was adopted as a stopgap crop to bridge the resulting hungry season gap.

This particular configuration of market economies seems to have been a colonial introduction, especially for the vast interior regions of the Gold Coast. British officials had a vested interest in integrating northerners into the markets, and the selling of subsistence crops was seen as critical to this and was strongly encouraged. As described by Tudhope in 1912, the ideal African subject was one who consumed manufactured goods and supported this consumption habit through selling produce at market, even if the profit margin was small (see also Guyer 2004). Africans have long been involved in market economies to various degrees, most especially at marketplaces, which have existed since precolonial times (Stahl 2018a). But there is less evidence for the existence of a free market—where prices were determined by competition and supply/demand. Multiple scholars have argued, in fact, that precolonial African trade operated according to a different set of principles, and its products were valued and accumulated very differently than in a supply/demand economy (Guyer 2004).

Market integration likely had profound social impacts that directly impacted food access. Watts (2013) argues that in Nigeria, the replacement of moral economies with capitalist ones focused on the individual was perhaps the most significant disruption of the colonial era. "Moral economy," a term popularized by James Scott (1977), refers to a precapitalist form that involved community mechanisms such as sharing and redistribution to ensure household reproduction for the majority. In other words, there was a community ethic that everybody had a moral right to access food. Social and political structures were therefore bound up in this "subsistence ethic," which provided welfare and insurance against food shortfalls in times of need (Scott 1977, 40). These social structures acted to minimize risk, particularly environmental variation, over the long term. We may juxtapose this with a market or capitalist economy that commoditizes labor as well as foodstuffs and other agricultural products. With this transition, the emphasis for smallholder farmers shifted from insuring community well-being to accumulating wealth as an individual or household unit.

Watts's well-known study emphasized the role of cash crop production in this transition process. As part of incorporation into the market economy, and under the direct influence of colonial governments, many people opted for or were coerced into cash crop production. Income generated by cash crops was then used to purchase the foodstuffs needed for household survival. This shift in focus, however, subjected small-scale farmers to the "horrors and moodiness of markets without the benefits of transformed forces of production. The tissues of the moral economy were stripped away, making peasants vulnerable to both market forces and a capricious climate" (Watts 2013 [1983], xxiii). Fewer scholars have focused on the selling of subsistence crops, a common strategy for farmers like those in Banda who could not produce lucrative cash crops. Not only was the profit margin smaller on these crops, but selling them at market directly removed these food-stuffs from the family larder. Food supplies ran out sooner than they might have previously, entrenching chronic seasonal hunger and vulnerability into the Banda region almost instantaneously.

The shift from moral to market economies meant that food was no longer a right but a commodity subject to supply/demand pricing. Throughout history, various leveling mechanisms have redistributed food to people in need, though imperfectly and unequally. Indeed, Thompson's (1971) original moral economy model was based on the history of how food prices responded to the removal of protective regulations and the emergence of a free market in Britain in the late eighteenth century. The free market for food was seen as a "natural" means to move food from areas with surplus to areas in need. Famines, of course, were possible for those with less ability to access cash, but in this age of Malthus this was seen as a "natural" check on burgeoning populations. Food prices soon were off the charts and led to widespread food riots. This transition—to food as a commodity on the free market—was catastrophic for many of the poor in Britain. Precolonial data are insufficient in Ghana to address when and how food became subject to supply/demand pricing, but it is likely that this change was associated with colonial economic policies as has been argued elsewhere (Mandala 2005; Watts 2013). Multiple sources suggest that this dynamic was at play in Taleland as well as near Banda in both the early and late colonial eras. Food prices at Kintampo, cited above, were double during the hungry season, as they were in Taleland (table 1). The need to purchase food through cash, especially during the height of the hungry season, would have spelled privation for a large number of poor people in the Gold Coast, much as it did in eighteenth century Britain. It is a shame that historical amnesia prevented British colonial officials from recalling this lesson.

In reality, the shift from moral to market economies is uneven, messy, and difficult to trace. "Moral economies," for example, did not lie completely outside the market in precolonial times. During the Atlantic trade, people responded to demands for craft production and trade on the Ghanaian coast by shifting their settlements to better maximize both (Kea 1982; DeCorse 2005). As Jane Guyer (2004) has demonstrated, the production of goods in order to access currencies has occurred for many centuries. African currencies have included cowrie shells, iron objects, and gold dust, among other things, traditions that extend back a millennium or more in some regions. In Banda, iron and copper currencies may have existed as early as the Ngre phase, and the presence of figurative "gold" weights as well as cowrie shells in the Kuulo phase suggests that such currencies were firmly embedded by that time. In other words, ideas of material wealth clearly existed alongside wealth in knowledge and people (Guyer 1995; Guyer and Belinga 1995). Maintenance of a moral economy may have helped accumulate wealth-in-people; assurances of welfare during times of need helped to ensure community well-being and underwrote the diverse assemblages of people and skills in a given community. Ideally, entitlements (land, cash, other tradeable goods) or access to food were thus ensured by community membership. However it would be a mistake to assume that "moral economies" (or Golden Ages) assured everyone equal or even adequate access to food.

Toby Green (2019) has argued that inequality became more pronounced in many West African societies in the centuries and decades preceding formal colonial rule, complicating a neat narrative of a shift from a moral to market economies. Through global trade connections in gold and human captives, some people and groups were able to acquire and accumulate wealth. This argument implies that the landscape inherited by colonial rulers was a profoundly unequal one, with major implications for food security. Add supply/demand food pricing to these socio-economic disparities, and the ingredients for persistent, chronic hunger among a poor underclass are clearly present. Colonial price records already show that food crops were double the price during the hungry season, speaking both to the high demand for these crops and their relative inaccessibility for people without entitlements. In Banda, we have little evidence for pronounced inequality among residents, but recall that integration into the Asante state laid the foundations for inequality on a larger scale by siphoning surpluses into state coffers. Reading Banda's record against this broader regional landscape of inequality illustrates the need to drill down into more specific local case studies in order to unravel the political dynamics at the heart of food access across space and time.

Many local societies retained or developed important social-leveling mechanisms that fought back against some of the problems associated with market economies. Some elements of precolonial moral economies, like food sharing, persist in places like Banda to this day in some forms even though they have been abandoned elsewhere (Cliggett 2005; Holtzmann 2009; Mandala 2005). By the time of Fortes and Fortes's visit, this process seems to have been well underway in Taleland. Fortes and Fortes (1936) detail how members of Tallensi extended households pooled labor, especially for agriculture. Working for the head of household may have been a way to access food in the early growing season, since labor is often repaid in food (Watts 2013, 127). But food was not shared among members of a household; most economic interactions were also independent. When food was needed, women would visit their extended relations in other villages or households. In fact, there do not appear to have been any nonkin redistribution systems to help the poor, as would be expected in a moral economy; instead, the chief was the richest man in town, and kept this wealth to himself and his family. Was this a pattern born of integration into market economies? The lack of precolonial data prevents us from evaluating this proposition.

In closing, it is important to highlight how the colonial government's presumption of what caused hunger limited its ability to address the problem, as this has implications for how we approach hunger in the continent today. Although nutrition in the interior regions was largely ignored, colonial officials did attempt to improve health through enforcing "sanitation" policies.[11] These policies were perhaps more consistent with colonial views of Africans as dirty and backward, and of themselves as civilizers. Changing nutritional outcomes would have meant recognizing malnutrition as a problem in the first place. This had potential to be a great embarrassment for the British colonial enterprise, particularly since nutrition went hand-in-hand with agricultural production, something that could not wholly be blamed on perceived African shortcomings. When nutritional problems were noted, such as Purcell's (1939) catalog of deficiency diseases on the coast, or Fortes and Fortes's (1936) description of the hungry season gap among the Tallensi, these were almost always blamed on African people and environments. While I agree with Shaffer (2017) that claims about the impact of colonial policies on nutrition need to be carefully evaluated on a case-by-case basis rather than generalizing about colonialism as a whole, it seems telling that the reverse explanations—that hunger has always existed—are rarely evaluated with the same caution. This "light touch" evaluation of colonialism also sidesteps clear evidence for the "malign neglect of officials" (Robins 2018, 168) towards scientifically documented severe malnutrition. Negligence may have been as destructive, if not more so, than targeted interventions, yet the major economic and social changes induced by colonial governments are let off the hook, because their effects were not deliberate.

Consuming a Remotely Global
Modernity in Recent Times

In 1957, the Gold Coast was the first African colony to achieve independence. It chose to call itself "Ghana," after the first of West Africa's fabled ancient empires. While the geographic boundaries of ancient and modern Ghana do not overlap, by adopting the name Ghana the modern nation-state sought to codify its first-comer status to independence, and to remind its inhabitants as well as onlookers of proud African achievements of the past (Gocking 2005). Sixty years later, strong nationalist sentiment remains in Ghana, but pride in the deep past has greatly diminished. One way in which the ambivalent and contested relationship between past and present is negotiated is through food preparation and consumption. Young people perform their modernity by consuming "modern" foods associated with urban dwellers instead of the "traditional" foods of their ancestors. The foods considered modern tend to be imported, processed foods that are part and parcel of the global food system. Over the course of the nineteenth and twentieth centuries, societies around the world have witnessed a dramatic rise in the consumption of industrial food products. Some societies have responded by placing greater value in local foods as a way to reassert cultural identity, but in others local food practices are associated with poverty and tradition and as a result are stripped of their cultural capital (chapter 6; Wilk 2006b).

In this chapter and the one to follow, I consider how—and if—we might bridge the gap between foods considered traditional and modern, local and global, and how we might connect the past and present more generally. In this chapter, I focus on the second half of the twentieth century and the early twenty-first century (c. 1940s–2009), a time when industrialized foods rapidly made inroads. Instead of providing a comprehensive overview of modern food practices, I focus on the tensions inherent in growing, making, and eating food in this "remotely global" (Piot 1999) context. I consider how and why people may actively maintain continuities in some food practices and quickly abandon others. This investigation reveals how food mediates and reinvents the relationship between so-called tradition and modernity. Through the constant need to navigate change and continuity, Banda's cooks have devised a food system that values flexibility and persistence at the same time.

TRADITION VERSUS MODERNITY? WHY DICHOTOMIES DON'T WORK IN WEST AFRICA

> Africa's . . . smallholder farmers toil in a time warp, living and working essentially as they did in the 1930s.
> —ROGER THUROW, *THE LAST HUNGER SEASON*, XIX

Some years ago, I was part of a group of faculty, students, and corporate employees who visited southern Africa as part of a food security internship program. The purpose of the trip was to expose the interns to the development arm of the corporation. We visited village after village that the corporation had helped, and in each we heard lines rehearsed for donor's ears. These visits were facilitated by partner organizations like USAID, whose representatives we also had brief opportunities to meet. It was in one such meeting that I first became aware of how pervasive the scarcity slot is in the development world, with real-life consequences for the men and women that are the subjects of their interventions. One USAID office was headed by a white South African who had fled black rule and and—somewhat ironically—was now in charge of "developing" other black Africans. So I should not have been surprised when he described "the Africans themselves" as the biggest challenge to development. As he put it, when faced with answers to their problems, like how to grow more or eat more nutritiously, sometimes they just didn't listen. When asked how he dealt with this problem, he threw up his hands and said, in effect, "Well, some people just don't want to be saved."

The USAID director was only expressing what passes for common knowledge about doing development in Africa, as captured by the quote that opens this section, from an award-winning monograph on African agricultural development by Roger Thurow. I am by no means alone in finding these kind of approaches to development wanting; anthropologists in particular have long been critical of development discourses in Africa (e.g., Ferguson 1990, 2006; Malkki 2015). These critiques question the taken-for-granted assumption that externally driven modernity and development are a good thing. They highlight how outsider expertise is privileged at the expense of African agencies and knowledges (see also Mitchell 2002). In my view, the long-term histories that are the focus of this book bolster these critiques by empirically interrogating the notion that African foodways are timeless and unchanging.

At the center of many of these stereotypes is the idea that Africans—and how they eat and grow their food—are nonmodern. Yet the African food history that I and others have presented conjures a much different narrative. As I have discussed in previous chapters, African foodways have long been intertwined in global happenings, from the Columbian Exchange detailed in chapter 2 to the Atlantic slave trade (chapter 3) and colonial rule (chapter 4). Few areas of the world lie outside of the global processes that have created modern economies and sensibilities, yet modernity has been assigned to some places and not to others. This is of course a politically and racially motivated project (e.g., Escobar 2012; Ferguson 2006; Pierre 2012). In a colloquial sense, modernity refers to specific developments in the West that were assumed to radiate to far-flung locations as part of the process of globalization. Anthropologists have argued that we need to provincialize that Eurocentric view (Chakrabarty 2000), and that how globalization has been defined, experienced, and created varies across local settings (Piot 1999). Modernity is often juxtaposed against tradition, though in practice the two are co-created (Hobsbawm and Ranger 1983). Modernity is commonly associated with a certain present progressiveness, with being modern, where tradition is associated with a backward past. The past is very often Othered as pre- or nonmodern, reminding those considered modern how far they have come (Cobb 2005).

Yet African modernities often have a different and more complex appearance than Western ones. Piot (1999) characterizes village modernity in his Togolese example as "remotely global." He shows how many apparently traditional features of village life, like the dependence on subsistence agriculture, are actually quite modern and the products of centuries of interaction with Europeans. Piot (1999, 2–5) laments how so-called African traditions are often villainized in news story imaginaries about the African continent, where they are equated with remoteness and inability to change. Challenges like food insecurity are blamed on African societies themselves rather than on the global modernities of which they are a part. These stereotypes plague many development initiatives on the continent, in which, as the USAID officer remarked above, the major hurdle is perceived to be Africans themselves. Yet as Piot (1999, 2) remarks, these kinds of ideas betray a remarkable ignorance of African history, one that borders on willful amnesia. They also turn a blind eye to the diverse array of strategies and improvisations that villagers must use on a day-to-day basis to sustain themselves and their families. Characterizing African societies as bounded, local phenomena means that the global, colonial, and postcolonial processes that underlay present-day insecurities go unacknowledged (Piot 1999, 17), which effectively focuses any solutions on the wrong unit of analysis.

In this chapter, I make a case for the remotely global nature of Banda foodways in recent decades, as a counter to the misconception of African foodways as traditional and unchanging that characterizes most development initiatives and even the low valuation accorded local foodways by many Banda youth (chapter 6). While there are many studies of food globalization, very few have focused on the African continent (but see Ham 2017; Holtzman 2009; Koenig 2006; Rock 2019). Like the Kabre, whom Piot (1999) describes, people in Banda have long adopted and experimented with nonlocal foods and goods, and documenting these processes has been at the center of the Banda Research Project (Stahl 1999b, 2001, 2002, 2007; see also bandathrutime.matrix.msu.edu and exhibits.library.uvic.ca /spotlight/iaff). These scholars resist the temptation to discuss African practices in terms of strict dichotomies like local/global and traditional/modern, as Holtzman (2009, 155–57) has cautioned against. To divide foods into such arbitrary categories means missing the ambivalences and ambiguities that largely define change and continuity in food practices. For example, Holtzman's ethnography of Samburu food describes the shift from a subsistence pastoralist diet to one increasingly focused on the consumption of purchased grains. This complex process involves a move away from more culturally salient and preferred foods like blood and milk and toward bowls of what is pejoratively described as "dry stick" maize porridge. Some Samburu see this as a sign of cultural decay while others see it as an example of progress, but ambivalence is often at the heart of these explanations. By refocusing his analysis on locally defined shifts in foodways, Holtzman is able to highlight the ambiguities at the core of remotely global foodways.

In the pages to follow, I explore some of the primary tensions in foodways over the last sixty or so years in Banda. This time frame is only an estimate. Holtzman (2009) shows the futility of imposing Western binaries onto local foodways and I discovered this many times over in my conversations with interlocutors in Banda. One of my first such lessons came when I asked an elderly female interlocutor how old she was. She laughed in my face, as did every other woman over the age of fifty.[1] Still, as an archaeologist obsessed with time I tried to establish a chronology of the changes people had seen in their foodways, through correlating a woman's life stage with certain major events like independence. While such a history would be useful for some, writing about shifting tensions rather than about linear events is more true to the experiences I recorded as well as to the nature of memory about food practices. To the best I can estimate, some of the earliest substantiated recollections in these interviews go back to the early 1940s, but most pertain to the last several decades. Most of my interview data was collected in 2009 and the tensions described in this chapter reference that point as "today," while acknowledging the limits of the present tense for capturing the rapid changes in food and farming that have continued to characterize the Banda area.

In the following section, I first consider how African food practices of this period have been relegated to a culinary underclass through discussion of Jack Goody's (1982) important work. Focusing on Banda next, I begin with the categories that local people deemed to be more stable, like seasonality, and move progressively toward the food practices that have seen the most change, ending with a discussion of food security. As will become clear, this scheme is inadequate to capture the dynamics of continuity and change, since even the most stable parts of the food system are constantly being reevaluated. Through discussing the tensions between choosing something new and continuing with something old, I try to bring local agencies and the constraints on choice to the fore. As a counter to the USAID officer's lament, I argue that it is the ability to make a choice, even if not under the conditions they would choose, that characterizes remotely global foodways in Banda. No wonder the "Africans" to whom he referred refuse to concede that ability.

COOKING, CUISINE, AND THE CREATION OF
A CULINARY UNDERCLASS

Jack Goody's (1982) *Cooking, Cuisine, and Class* has long been considered a path-breaking work on the anthropology of food, particularly in Africa (Mintz and Du Bois 2002), and looms large in any discussion of African foodways. Following the theoretical conventions of his generation, Goody's central project was comparative, seeking to understand how and why his African case studies differed from what was known about food history in the rest of the world. He compared the foodways of the LoDagaa and Gonja of northern Ghana to those of Asian and European civilizations, and argued that African cuisines were largely undifferentiated in comparison. Specifically, Goody argued that Africa lacked *haute* cuisines, since he found no evidence of the specialized chefs, techniques, or ingredients that defined the elite cuisines in Asia and Europe. He explained the absence of African *haute* cuisines through a corresponding lack of strongly stratified social classes.

Goody's structural analyses are problematic on several fronts, though his high-quality ethnographic data continue to provide insight. One of the most obvious critiques is his use of an essentially European and Asian yardstick to evaluate African foodways. While he did not intend to denigrate African cuisines, some scholars argue that his work is at least partly to blame for an avoidance of African food as a site of scholarly inquiry (Lyons 2007, 348). Goody's central project had the unintended effect of Othering African food practices in ways that continue to be commonplace among outsiders. African foods are not readily appreciated by most Western palates (Lyons 2007, 348), for reasons I explore and this and the next chapter, leading scholars and foodies alike to ignore their culinary, cultural, and nutritional values. As I detail below, this plays out in all sorts of forms, from describing African foods with pejorative language to treating African foodways as one undifferentiated mass. Part of the problem is that African taste preferences are realized and expressed in different ways than Western ones. While Westerners value novelty, many West Africans value quantity, a point raised by Goody (1982, 67–68) and expanded by De Garine (1997).

One consistently overlooked quality is texture, of which people in Banda and elsewhere in West Africa are exceptionally discriminating consumers. The importance of texture is readily apparent to the ethnographer interested in food preparation, even with an unskilled Western palate, because creating the right texture takes tremendous skill and time on the part of the (usually female) cook. Given his fascination with *haute* cuisine, which is by and large discriminated on the basis of specialized preparation, I was surprised to discover that Goody (1982) devoted very little of his energies to elucidating African food preparation practices. In his chapter on LoDagaa and Gonja foodways, preparation receives less than three pages of coverage, while production and consumption receive over twenty pages each. This may simply reflect his positionality as a man, but it also may stem from his earlier interest in surplus production as the root of differences in Africa versus the rest of the world (Goody 1977; Goody 1982, 58–61).

One last critique needs to be made, and that is that Goody's structural and comparative framework does not wholly match the data he collected. His book is based on thirty years (1949–79) of repeat visits to Ghana during the transition from colonial to independent government. This makes his work an excellent comparison point for my investigation of changing foodways in Banda over a similar time frame. What is odd about his resulting work is that although change is frequently mentioned, he does not structure his analysis around those changes, instead focusing on the structural differences between African and Eurasian societies, which are less meaningful for understanding African foodways as a whole.

These critiques suggest that Goody's work was very much a product of its time and as a consequence many of his key positions need to be reevaluated. Fortunately, he presents a great deal of high-quality ethnographic data and observations, which I've used as points of comparison to Banda, particularly since they cover a similar stretch of time and correspond to societies that are geographically proximate to Banda. In the pages to follow, I tack back and forth between Banda's foodways and those of the LoDagaa and Gonja as described by Goody, but expand the focus to food preparation and women's work.

THE SEASONALITY OF CUISINE

When I first approached men and women in Banda to talk about changes they had seen in food practices over their lifetimes, the most common response I received was that there had been no change. At first, I was more interested in changes than in continuities, but as the global food literature has pointed out, continuity too is agential, and is often actively protected and promoted as a means to deal with external pressures (Wilk 2006a, 2006b). In this and subsequent sections, I begin by talking about elements of cuisine that are considered more stable before moving to some of the major changes in Banda's culinary landscape. As I will detail in coming pages, continuity is characterized by a remarkable degree of fluidity, one that actively incorporates considerable change while retaining the overall shape of food practices.

Foods have seasons, as my interlocutors often said, and in Banda there are two major seasons that dominate the agricultural and culinary landscape: the wet season, when the most rainfall is received, corresponds with the growing season, whereas the dry season is characterized by the coming of dry Harmattan winds from the Sahara. In practice, farmers in Banda recognize seven seasons that take into account varying amounts of rainfall and suggest when it is best to plant certain crops (figure 5; Logan 2012, 51–56). People used to rely on two major staples that are differentially available based on the season. Yam *fufu* (*gbosro* or simply *sro* [literally, food] in Nafaanra) is the primary food from the beginning of the yam harvest, sometime in August or September, until they run short. *Tuo zafi* (Hausa for "hot porridge"; *kambɔ* in Nafaanra), or TZ, as it is known in Ghanaian English, is a thick, firm preparation of a polenta-like consistency made from grain and cassava flour. It dominates when yams run out and is eaten throughout the wet season. Other staples from southern urban areas, like *kenkey* and *banku*, both variations of fermented maize dough, have also gained in popularity.

Soups (*chiin* in Nafaanra) add flavor and texture to fufu and TZ and also utilize seasonally available plant and animal foods. Yams are often complemented by soups made from ingredients available in the late wet and early dry season, which during the wetter parts includes wild greens, freshly harvested *fnumu* (squash seeds) or groundnuts (peanuts), as well as by tomato-based "light" soups made from fresh tomatoes, garden eggs (small eggplants), and *okro* (okra).[2] As the weather dries and these ingredients become more scarce, people make soups out of dried (e.g., leaves and *okro*) or purchased (e.g., canned tomato) alternatives. Other purchased alternatives include palm nuts from the wetter south, which are made into a rich, velvety palm nut soup. When the wet season returns, people complement TZ with soup made from fresh *okro* and other vegetables, including a wide range of cultivated and wild leafy greens. Meat or fish in small quantities are a frequent ingredient in all soup varieties; the soup is considered "raw" and not very nutritious without their presence.

Within the confines of seasonal food availability, people enact a number of taste and texture preferences. Chief among them is the need to consume soup with a "slippery" texture as an accompaniment to *tuo zafi*, but not to *fufu*. A slippery soup acts as a vehicle that guides the *tuo zafi* down one's throat, meaning there is no need for chewing first, since, as people told me, the "mortar has done the chewing for you." As Richards ([1939] 1995, 39) recounted among the Bemba, this desire for slipperiness may relate to the (formerly) gritty nature of the starch, which in times past included extraneous matter from grinding stones. *Okro* is chief among the mucilaginous vegetables, but women also use a selected array of wild leaves that impart the same textural quality. Both *okro* and one wild leaf, *lom*, are dried so that this texture can be produced in the dry season. People also extend the seasonal availability of some other products by purchasing processed alternatives like canned tomatoes or sardines.

At the time this research took place, in 2009, most farmers focused on growing enough subsistence crops "to eat and to sell." This is a low-risk strategy, since it insures that food is available from one's farm in most cases, though as discussed in chapter 4, it also has some disadvantages. Besides staple carbohydrate crops

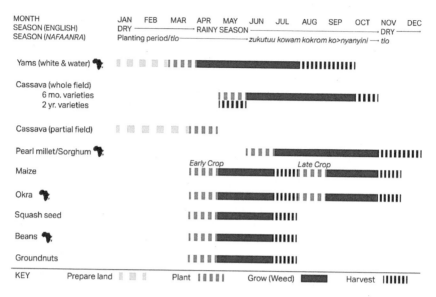

FIGURE 5. Agricultural cycle in the Banda area, 2009. Crops marked with an Africa symbol are indigenous to the continent.

like yams, cassava, sorghum, and maize, other crops in this eat-and-sell category include peanuts, *fnumu*, and beans (mostly cowpeas). Vegetables including *okro*, garden eggs, peppers, and more are also grown on a smaller scale. Many farmers also grow cash crops, particularly if they have access to additional land or to farming inputs like fertilizers and pesticides. Cash-cropping likely has a long history in Banda, extending into the colonial era and including crops like cotton, peanuts, tiger nuts, and tobacco. In 2009, farmers tended to focus on teak, which took many years to grow but had potentially significant returns, as well as cashew. Ann Stahl (personal communication, 12/2018) reports a significant shift to cashew in the last decade, even at the expense of yams, a highly prized food.

Although seasons do provide a natural rhythm to people's food routines, seasonal food availability is not separate from global forces but is shaped in a considerable way by climate change as well as by market integration. A shift in rainfall regimes is one area that people have little control over, one that is having major impacts on what people grow and eat. Farmers complained frequently to me of erratic rainfall, with some years receiving too much rain and some years too little, as well as of increasing unpredictability regarding the onset of the rainy season. Meteorological data from Kumasi confirms these complaints, showing a reduction in rainfall beginning with the Sahel drought of the late 1960s-early 1970s, which marked the beginning of a drought that continues to this day (figure 1; Shanahan et al. 2009). The timing of the start of the rains is particularly problematic and can upset the entire agricultural calendar (figure 5). A late onset, for example, means a shorter growing season and smaller or less productive crops overall. General unpredictability means farmers are unsure as to when to plant their crops in order to take full advantage of the rains. As one farmer put it, the amount and duration of the rainy season changes, but crops still need the same pattern of rainfall. If the rains do not cooperate, harvests and the people who depend on them suffer. Some farmers noted that in bad years there was barely enough to harvest, and certainly not enough to sell.

Changing rainfall is not the only pressure on Banda farmers; most farmers noted that another significant change was the need to produce a greater quantity of crops. My interlocutors described this as a significant shift since the days of their parents, when subsistence farming was the norm. Increased demand for cash meant that they were now pressed to produce enough to eat and to sell at market. The pressure to produce an increased quantity, particularly in the face of declining and abnormal rainfall regimes, has had several negative effects. Several farmers mentioned increased pressure on the land due to the need to farm more. Farmers could intensify production through the application of fertilizers, but this input was deemed too expensive for most. Instead, many people have adopted an extensification strategy, farming larger or more parcels of land in order to produce more. Although Banda's low population in the last century has meant that land has usually been abundant, that is starting to change as people try to keep up with changing demands for cash. One notable impact is that people are reducing the duration of fallows on land, with negative impacts on soil nutrients. While many farmers described a ten- to fifteen-year fallow as ideal for growing yams, most remarked that they often shortened this to five years, the minimum number needed to grow cassava and guinea corn.

As of 2009, with the exception of plots with cashew or teak trees, parcels of land are generally farmed for two years before being left to fallow. To prepare a field for planting yams, grasses are burned and non-useful trees are removed. Fruit trees and other useful trees are left in the fields. Other small trees or branches are often left in place or are added to help stake the growing yam vines. Mounds are raised after a little rain when it is easier to dig, though when this rain will fall is very unpredictable of late. Field preparation typically occurs in the very end of the previous wet season or the beginning of the dry season. In a first-rotation field, yams are the first crop planted in the mounds, sometimes as early as November, but generally sometime between January and April. Freshly planted yam tubers are often covered with spare vegetation and dirt to protect them from the dry-season sun until the rains start. As soon as yam planting is finished, other creeping crops like calabash or cowpeas can be planted in the mounds, taking care not to plant those that will quickly grow and overshadow the fledgling yam vines (figure 6). *Okro* can also be planted in the mound at this time, but is not usually planted in quantity as it is produced primarily for local consumption. Later, maize is often also planted in the sides of the mounds. Cassava may be planted along the edges of the fields

FIGURE 6. Schematic cross-section of mixed farming on mounds.

in rows; since it is deep rooted, cassava is not planted in the same mounds as yam so as to avoid competition for the same soil nutrients. After these crops are planted, the farmer weeds or waits for the crops to sprout and then weeds the field. Ideally, this portion of the agricultural cycle occurs in the late dry season and early wet season, so that by the time planting is finished there has been adequate rainfall for both; however with the changing timing of the rainy season's onset, planting may occur too early or too late to benefit from the rains.

In this first rotation, staple crops are grown on mounds that function as micro-environments providing a niche for each crop. This ingenious system takes advantage of the different depth and soil requirements of each crop (figure 6). Yam and cassava require deep, loose, well-drained soil, which the mound provides. This leaves the top layers of soil free for shallow-rooted crops like beans and *okro*. Timing is also critical; all of the crops grown in the top layer of the mounds mature in four months, compared to six months for yams and six to twenty-four months for cassava. This insures that the crops in the top layer do not interfere with yam growth, and the yam plant is not yet very large at the time they are harvested. Yams or cassava are then left in the field to mature; yams are generally ready in August or September. Cassava can be harvested at the same time in the case of six-month varieties; others remain in the field for another six to twelve months to reach optimal yield. In this case the crop may occupy the field for the duration of the second rotation. Because cassava remains in the field for so long, it can be harvested year-round if needed, although the tubers may not have reached maturity.

In the second year of cultivation, as soon as the yams are harvested, the land is cleared. After the rains begin, ridges are raised to plant *fnumu* or groundnuts. Sorghum (or millet in the past) is planted in June when the rains are most predictable. It is ready to harvest six months later. Maize can also be planted at this time, but this can be a costly option if it was planted in the previous rotation because of the need for chemical fertilizers. For this reason, sorghum is preferred in the second year. It requires only phosphorous, generally produces good yields, and can easily be sold. Weeding must be regular to ensure that spear grass, an introduced weed, does not compete with grain crops for nutrients. By year three, spear grass begins to dominate and most farmers do not find it worth their time and energy to cultivate another year in light of declining soil fertility.

Monocrops of maize are also planted, though they are not as common as mound plantings. The reasoning is that because yams are highly valued for eating and for sale, it is a waste to use newly cleared fields just for maize. Maize can be ready in as little as three months, though it has not yet dried, making harvest more difficult. To ensure planting of two maize crops in one growing season, farmers can plant the first crop in April, anticipating a June yield; it can be left to dry for a few weeks to a month, which is preferable. Another maize crop is then planted in August, to be ready in November, before the other grain crops, at a time when yam stocks run low.

Both men and women do farm work, though in recent times there has been a shift to women having their own farms. This allows women a steady supply of income that is under their control, and also evens out risk of crop failure for the family. Traditionally, in both Banda and the areas studied by Goody (1982, 56), men did the heavy farming work such as clearing, weeding, planting, and removing plants from the ground at harvest time. Women assisted by weeding and tending, peeling and drying, and in final harvest practices such as gathering grain heads and processing them into clean grain. Yams were associated with men, who were responsible for growing them; women assisted by carrying seed yams to a new farm for planting and grown yams from fields to kitchens. These gender roles have changed as more and more women run their own farms and often produce lucrative yam crops. The most labor-consuming tasks are clearing of the fields, building mounds, and weeding. In recent times, both men and women have started to hire Dagaarti (LoDagaa) laborers from northern Ghana—the very same group Goody studied—to perform these tasks.

The hiring of farm labor is a relatively new development in central Ghana that is linked with the commodification of both food produce and labor. As Goody (1982, 73, 88–91) details, men used to work first on the farms of the senior member of their household, then on their own. Laborers, both relatives and not, were paid in beer and sometimes food, with the expectation that the host would reciprocate with his own labor and that of his kin in the future. My Banda interlocutors mentioned a similar system in Banda. This system worked because the unit of production loosely matched the unit of everyday consumption. Those who farmed together also shared prepared food and the same granaries. Even by the time of Goody's writing, this system was beginning to fray, and indeed in Banda it is no longer the norm. The unit of production and consumption has become the conjugal family rather than extended household, though in practice food is often shared to some members of the latter. Goody (1982, 88–91) blames this shift on integration into market-based economies, in which labor and its produce have cash value. In these settings it becomes difficult to engage in reciprocal labor relationships.

In order to deal with increasing demands for cash and changing rains, people have increasingly altered what they eat. In general, the crops that fetch the highest price at market are consumed less and less within farmers' homes, and are instead sold to purchase less-expensive staples. Cassava in particular is being consumed more and more. It is one of the most commonly farmed staples in the Banda area, and is cheaply and readily available throughout the year. While the tuber is nutritionally lacking, the leaves are full of vitamins and nutrients and are often made into an accompanying soup. Yams, as the most valued staple of the area, were still eaten at the time of my research, especially during the yam season in 2009, but it was more profitable to sell one's harvest and eat cassava instead. A recent intensification of cashew farming instead of yams means that people are now eating even more cassava (Ann Stahl, personal communication, 2018).

Farming while remotely global means producing crops for market. As I will discuss below in the case of pearl millet, this means some crops are abandoned because they fetch lower prices at market, while others that fetch higher prices, like yams, are sold instead of consumed locally. The proceeds are often used to buy the cheapest foods available rather than the most nutritious or desirable; this is a strategy that actively balances food preference with the need to acquire cash. Agricultural development dollars tend to promote research and expansion of a narrow range of subsistence crops, particularly maize and rice, which also has down-the-line impacts on food availability. Food practices in places like Banda are thus whittled into a form and shape that make sense given market constraints.

"*TUO ZAFI* IS ALWAYS THERE FOR YOU": A FLEXIBLE CONSERVATISM

If there is one constant in the African continent, it is change. Practices which retain some semblance of themselves over time have to allow for a high degree of flexibility in order to persist. People may readily adopt new ingredients but continue to use the same food preparation techniques and meal structures (Dietler 2007, 224). *Tuo zafi* or *kambɔ* is a starchy staple that displays remarkable flexibility and staying power. It can be made from maize, cassava, pearl millet, sorghum, and probably any other starch that can be converted into a flour. To make this salient for Western audiences, consider the European mainstay of bread, which can achieve the proper texture using only a very narrow range of gluten-rich grains, and is nearly always made with just one—wheat.

People allow for this flexibility in ingredients by adhering to a very rigid and lengthy set of preparation steps (figure 7) that can transform virtually any starchy crop into a culturally recognizable product. To make *tuo zafi*, women must first produce a flour out of whatever staples they have available. In the olden days, women would have employed a grinding stone or wooden mortar, but today the overwhelming majority of women pay to have their starches ground at a diesel mill (see below). Flour from the mill is sieved to remove any chaff or large chunks, to insure that a desirable fine texture is achieved.[3] Women add a bit of flour, usually maize or in the old days pearl millet, to a pot of boiling water to make a thin porridge. Once thickened, handfuls of maize, millet, cassava, or sorghum flour are

FIGURE 7. How to make *tuo zafi*.

added to the porridge, and rapidly stirred. As the mixture thickens, women adopt a special stroke, illustrated in figure 7, which beats the *tuo zafi* into a very firm yet fine mass. The paddling is so vigorous the cook must secure the pot with her feet. In the past the process could easily result in broken earthenware pots, which is one of the reasons women were happy to switch to their aluminum counterparts.

Making *tuo zafi* is not easy. It requires considerable skill, strength, and patience. The whole process takes one to two hours with already ground grain from the mill; that time doubles if one has to grind or pound grain oneself. Women I talked to who were from outside of the region said it took several years to learn the requisite strokes and methods to make passable TZ; indeed my attempts to participate in making *tuo zafi* resulted in a sad, soft mess that little resembled the hard, springy masses people are used to and expect (results visible in figure 7). Making *tuo zafi* is a deeply embodied skill, one which young girls are often trained in starting in early adolescence, and are encouraged to make on their own only much later.

These narratives, which speak to the skill required to produce passable *tuo zafi*, also known in the wider region as *tô*, contrast markedly with common perceptions about the food by Westerners, who tend to regard it as a dull, monotonous substance. The most commonly used English translation of the food—simply "porridge" (e.g., Goody 1982)—does it no favors in this regard. Calling TZ porridge equates it with a tasteless gruel that takes little time or skill to prepare and does not account for the springy, firm texture that so many African consumers recognize and prefer. It is but one example of how African food is imbued with negative connotations, a tendency common to many formerly colonized areas (Janer 2007).

The benefits of being able to produce a culturally acceptable food from virtually any starch are manifold. It means new and nonlocal starches can readily be adopted, as we might imagine maize having been at its first introduction in the early seventeenth century. But it also means that if a preferred staple is unavailable, using an alternative one may not be perceived as quite as much of a hardship, because you are still eating a preferred local dish. That these capabilities are desired is clear in the persistence of *tuo zafi* throughout the twentieth century, though quickly produced foods like rice are increasingly available. While this continuity may relate to availability of funds to purchase imported foods, *tuo zafi* also has a staying power that speaks to how people have navigated the tensions between local and nonlocal foods as well as the varying availability of different staples.

"OUT OF NEED": WILD LEAVES, HISTORY, AND THE POWER OF PERSISTENCE

The quintessential Nafana meal is *tuo zafi* accompanied by leaf soup (*chiin*). Leaf soup demonstrates many of the same qualities as TZ, but allows for a much greater degree of improvisation and creativity. The method of preparing leaf soup is fairly straightforward, since the primary goal is to soften the leaves of wild, collected, or cultivated species. This is accomplished through pounding, adding quake lime or ash from certain plants, and boiling, sometimes more than once. These methods can reduce almost any edible leaf into a soup. Flavorings and textures vary according to the staple served and the cook's preference.

But compared to TZ, which is usually made from the three commonly available starchy staples, leaf soup can be made from over sixty different wild leaves, as well as from the leaves of many common crops (predominantly cassava and cowpea).[4] In some cases, the leaves are mixed with ingredients like garden eggs (*piɛwɛ*), *fnumu* (squash seeds), and groundnuts (*bongrɛ*) to add texture, flavor, and nutrients. Additional flavoring agents may also be added, including meat or fish, wild mushrooms, fermented seeds (*dawadawa*, kapok, baobab, or cotton), or in more recent days, Maggi cubes (see below). A number of fresh wild leaves may be selected because they impart a desirable mucilaginous texture, often denigrated by outsiders as slimy. Finally, several types of leaves (especially the wild basils, *napua* and *chasigbɔɔ*) are added as flavorants themselves.

Leaves are obtained from wild, weedy, semi-cultivated, and domesticated species. At the time of my interviews, the majority were common weeds of cultivated or residential areas, many of which were also purposefully planted or encouraged in gardens and near villages. While the wide array of leaves used might be taken as an indication that women use everything in sight, this is not the case. Many taxa are toxic, and those that are used are clearly selected for certain qualities like texture or taste. The wide but selective range of taxa used demonstrates a detailed understanding of local environments as well as of plant lifecycles, which stands in contrast to the stereotype that Africans have little control over nature (see introduction). Collecting and nurturing the appropriate species requires substantial knowledge and time and represents complex people-plant relationships.

The use of wild leaves in soup is also a historical record of conflict and survival. Leaves figure quite prominently in oral histories of wars in the nineteenth century (chapter 3). This was a time of raiding and warring between the Banda peoples and neighboring groups, and the Banda were sometimes forced to flee from their villages at a moment's notice. Oral histories suggest that instability persisted periodically over decades, making farming on a regular basis impossible. People recount stories of their ancestors eating leaves from the bush: one person would try an unfamiliar leaf, and if they didn't die, the leaf was safe to eat. Hunted animals, along with cassava plucked from fields they were running through, are the only other foods associated with this time of upheaval. Oral histories, especially ones that recount such a dramatic time, are prone to exaggeration, but linguistic evidence suggests that leaves were a component of diet in times of need. The literal translation of the names of at least two wild leaves (*kalameshia, kemeshiama*), both associated with this period but no longer in use, is "out of need."

The use of wild leaves was repeatedly mentioned in interviews as a strategy to cope with seasonal and severe food shortage. The most severe shortage in living memory occurred in the early 1980s, in association with both a severe drought that affected much of West Africa and political upheaval in Ghana. Almost everybody interviewed said that during this time, people simply ate leaf soup and slept, since the starchy staples that usually accompanied the soup were in short supply. People also turned again to cassava, and the most unfortunate were forced to eat it without detoxifying it first, resulting in stomach complaints. Although many people today make use of leaf soup as both a regular and shortage food, few remembered the association with the nineteenth-century wars. This pattern of selective forgetting may index the ways that people remember past events via bodily knowledge. Recent work on food and memory has stressed the importance of bodily actions, repetition, and the senses in how people remember food and the events associated with it (Counihan 2004; Holtzman 2006; Sutton 2001). The bodily practices that people perform routinely—like cooking wild leaf soup or *tuo zafi*—may be resistant to change in part because many become subconscious once the skills are initially acquired. These practices are learned from one generation and passed to the next, much like other crafts (e.g., pottery production), forming bodily archives of food history (see also communities of practice, e.g., Roddick and Stahl 2016; Wenger 1998).

Conservatism in food preparation practices may play an important role in navigating change and insuring food security. The preparation of both TZ and leaf soup displays remarkable flexibility and persistence. This conservatism—but at the same time, flexibility—in food preparation methods ensures food security because of the great range of plants and animals that can be used to produce the same culturally acceptable meals. Recall that modern definitions of food security entail that all people have enough food, can access it, and feel secure about their food supply (Maxwell 2001)—and that part of access and security is the cultural acceptability of any given food and the knowledge about how to prepare it. Food preparation methods in Ghana and West Africa more broadly are quite complicated and time consuming. Such complicated operational sequences serve the dual purpose of producing a culturally desirable product and of ensuring that any number of a wide variety of ingredients can be used. Flexibility is particularly important when dealing with unpredictable environments and economies, and frequent seasonal shortages that characterize much of the broader West African region today.

Persistence may also play a role in dealing with rare but severe food shortages, such as the 1982–83 drought. Minc (1986) has suggested that, in the Arctic, myth and ritual act to reinforce and remind people of coping mechanisms in times of both common and rare, severe shortages. Because leaf soup is the soup most closely identified with Nafana culture, we can conjecture that its social importance may help it to serve a similar function. Perhaps we can think of conservative food practices and preferences themselves as a way of preventing and dealing with food scarcity. As Wilk (2006a) has pointed out, keeping things the same takes work; conservatism in food practice might be best described as an actively maintained cultural value rather than a stubborn holdover from the preindustrial past.

Yet even the most conservative food practices are open to negotiation when they collide with opposing priorities, especially in remotely global societies like Banda. Some women observed that people were increasingly using the leaves of crops (especially bean leaves [*susuwere*] and cassava leaves [*dwawere*]) rather that collecting the wild and semi-wild herbs that account for much of the variation in daily diet. This is in part an environmental issue, as rising use of herbicides has decimated these plant populations and increased the amount of time women must spend searching for them. Collecting wild leaves is a laborious task in a context where women are under increasing pressure to generate cash income, which forms the subject of the next section.

MAGGI: CONGEALED CUBES OF WOMEN'S LABOR

To truly appreciate the work that has gone into passing down knowledge about wild leaves, as well as the time it takes to collect them, we have to consider the many pressures on women's labor.[5] In this section I focus on how women have selectively adopted some food preparation technologies and ingredients that free up their time for other pursuits. In particular, I will focus on the uptake of imported and industrialized foods as well as use of the diesel-powered grinding mill, because they reveal the ambiguities and tradeoffs involved in adopting new foods and preparation methods.

The spread of industrialized foods over the last century has had lasting and significant impacts on both nutrition and women's labor all over the world. This food globalization has been associated with declining health (Popkin, Adair, and Ng 2013) and the erosion of local food traditions. Although viewed somewhat pejoratively by Western food movements today, it is important to acknowledge that these new processed foods also afford a host of opportunities. Industrial foods involve the processing and storage of ingredients into a more shelf-stable form, effectively lengthening the seasonal availability of some foods. Many African consumers consider these foods the most hygienic and pure, since they are produced in modern factories and are less susceptible to adulteration than home-prepared foods. But perhaps the chief attraction of these foods is convenience: cooks can simply pop open a can of tomatoes or sardines instead of preparing them from scratch (Goody 1982, 180).

In contrast to the West, where industrialized foods are today viewed with suspicion and disdain, in places like Banda industrial and imported foods in general tend to possess a higher cultural value than many local alternatives because in addition to alleviating the concerns mentioned above, they are associated with modern, urban living. This is not a new pattern; as discussed in the last chapter, the association of imported foods with status was commonplace among educated Africans in the colonial period (Goody 1982; Robins 2018). Today, there is also a general belief in Ghana that rural, village life is less desirable because it is less developed and therefore backwards. Most men and women who now reside in the Banda region have spent some time in an urban area, seeking out wage-labor opportunities or furthering their educations. These experiences expose young people to urban food habits, to which they aspire once returning to the village (chapter 6).

Several processed and imported food products have been adopted in Banda over the last three decades, though their use varies with economic class. Bread baked from imported wheat is not made in Banda, but can be found in Wenchi (the nearest large market town), along with canned milk, tinned coffee, and tea. These foods are uncommon among all but the most well-off individuals in Banda but are considered highly desirable. Whenever I was heading into the city, people requested loaves of bread from specific towns, since they appreciated the different recipes as well as the novelty of a food unavailable in Banda itself. While these requests could be interpreted as a sign that Banda residents valued novelty, my other experiments in that regard suggest this is not the case. Instead, there is a marked appreciation of foods from specific regions, including bread and various wild plants and animals, suggesting a developed sense of *terroir* or taste of place (Trubek 2009).

Asian rice is also considered expensive relative to other starchy staples, but is strongly preferred among younger women and is growing in usage. Rice is the mainstay grain of urban dwellers, which partially accounts for its higher status; it is also very quick to prepare compared to local starch dishes like *fufu* and *tuo zafi*. Fresh or frozen meat is highly sought after and is considered the ingredient that makes a soup "strong," or healthy, but it is consumed in very small amounts. Fish, usually in dried or canned form (e.g., Teacher brand mackerel), is cheaper and consumed more often. More commonly purchased are cheap soup ingredients, which also save women a considerable amount of labor. Canned tomato paste can be combined with vegetables to produce a quick light soup, which has become very popular in Ghana, enough to support the growth of Ghanaian tomato-canning factories.

The use of Maggi cubes—a Nestlé bouillon product most often found in shrimp flavor—captures the tensions at the heart of Banda's industrializing foodscape. Maggi is the ultimate industrial food success story in the African continent, in part because of its availability to all but the very poorest people. In northern Ghana, Ham (2017) describes ambivalences about Maggi's use as compared to locally available *dawadawa*. *Dawadawa* is a paste made from the fermented and ground beans of a wild leguminous tree (*Parkia biglobosa*) that is used as a flavor enhancer and nutritional booster in soups in Banda as well. But *dawadawa's* pungency gives it a strong smell that is much eschewed by urbanites, so much so that the women Ham interviewed described it as nonmodern. Maggi, on the other hand, provides a neutral smell and flavor that are deemed more widely acceptable. Yet a large majority of women still used *dawadawa* in their soups, because they recognize its nutritional qualities. *Parkia biglobosa* beans contain protein, which is enhanced through the fermentation process, while Maggi is almost pure sodium (although Nestlé has recently announced the addition of iron supplements for African markets). The fact that Maggi has not replaced *dawadawa* in Ham's study area is all the more surprising when one considers the tremendous amount of labor required to produce *dawadawa* paste from the raw beans, as compared to the simple unwrapping of a bouillon cube.

In Banda, the use of *dawadawa* persists to a limited extent alongside Maggi. *Dawadawa* was most commonly referenced when talking about *wehan* (figure 8), a steamed dish made during times of food shortage from ground *Parkia biglobosa* beans and cassava scraps. Many other flavoring agents have been abandoned. Seeds from baobab, kapok (*kondo*), and cotton (*kombotoo*) can be processed much in the same way as *dawadawa*, through grinding and fermentation, to produce flavorful additions to soups that help to substitute for meat. Wild mushrooms were also in widespread use in the past, but are no longer commonly used. The near abandonment of these alternatives is no surprise given the labor required to produce them. Maggi cubes in many ways represent women's labor, sacrificing distinctive flavor and enhanced nutrition for time spent in the kitchen.

Maggi's popularity may also correspond with a marked decrease in meat availability in Banda. People used to access a wide array of bush meat on a regular basis, but in recent years the supply of bush meat has declined rapidly. My interlocutors suggest that two forces are at play. The first is the enclosure of Bui National

FIGURE 8. *Wehan*: steamed *dawadawa* and cassava, often eaten with a sauce (see table 2).

Park, which runs the length of the Banda Hills and expands westwards. Hunting is not allowed in the park, where "poaching" is deemed a punishable offense. Many people remarked that animals sought refuge in the park, causing access to bush meat to decline. Others related the lower availability of bush meat to overhunting in order to profit from the sale of the meat. Given Banda's long histories of wild animal consumption (Logan and Stahl 2017; Stahl 1999b), this is a significant departure from past culinary practice, and likely one with significant nutritional consequences. Maggi fills this gap in terms of flavor, but not in terms of nutrition. Increased use of dried and canned fish has also helped make up for the shortfall.

Whether it motivated the adoption of Maggi or was an unforeseen effect of it, how woman use their time has changed dramatically with new food technologies (Logan and Cruz 2014). But as Ruth Cowan's (1983) landmark study on the introduction of kitchen technologies in the United States cautions, new technologies (e.g., microwaves) do not actually ease women's share of labor within family units. Instead, women tend to fill those spaces with other productive, though often not remunerated, tasks (DeVault 1991). In Banda, there is a greater emphasis on remunerative activities than DeVault observed in her Chicago-based study. One of the most significant changes has been initiated by the introduction of the diesel-powered grinding mill, which has replaced the use of grinding stones and the need to pound grains or tubers into flour. Grinding or pounding enough grain for a single day's supply of flour used to take about two hours. Today, most women make use of the mill unless they are out of money to use it. But when mills first started arriving in Banda in the 1980s, women met them with mixed reactions. Some were hesitant due to the cost; others claimed that mills produced an inferior product that spoiled more quickly. But most seem to have welcomed the mill because of the time it freed up. Ama Benimbedei (approximately seventy-four years old), from Nyiire, told me that before the mill, women had to wake up very early, at around 4 a.m., to pound grain before heading to farm at 6 a.m. The mill freed up time, so that women could start to make soup earlier or go to farm earlier, or could sleep a bit more. Others, like Abena Sarabi (approximately sixty-five years old) recall devoting the extra time to activities that could generate cash, like soap-making or collecting kapok fibers for sale. Some women, including Ama Benimbedei, also brought up the unintended social impact of the mill. Many young women would gather in the early mornings to pound grain together. Elderly women talk about this as a fun gathering with songs, jokes, and gossip. They regretted the loss of this gathering time, and pointed out that women no long gather in groups on a regular basis, especially with the erosion of nubility rites (chapter 6).

The increasing use of industrialized foods as well as of grinding mills have also more fully incorporated women in particular into cash-based economies— apparently by choice. Buying these goods and services means that women save labor, but in order to access cash, women must invest that labor in productive activities, which begs Cowan's question of whether or not technologies actually save women time or worry at the end of the day. The need to access one's own cash stream in order to buy food has also resulted, for some women, in increased independence from their husbands. While women have long helped out in their husbands' fields, and even cultivated their own cash crops like tiger nuts, younger women are increasingly taking control of their own farms and growing the same crops as men. For some women this enables them access and control over their own finances. For others, who share finances with their husband, having their own farms is a risk-averse strategy. This is a major break with more gender-segregated agricultural tasks of the past, but men and women claim that it helps insure a harvest in the bad years. The skill of the farmers differs, and the added variable of cultivating different plots of land introduces a degree of randomness that helps deal with the often unpredictable microenvironments of the Banda area.

PEARL MILLET, FOOD OF THE ANCESTORS

Negotiating labor dynamics in an increasingly monetized remotely global economy has meant the reevaluation of what is grown, even in the case of a culturally valued staple grain like pearl millet. As I have shown through this book, pearl millet formed the mainstay of Banda diets for much of the last millennium. Pearl millet had tremendous cultural value and also afforded people a degree of resilience to major environmental changes, particularly drought. Although not the highest-yielding choice, pearl millet was probably the best risk-reducing crop that Banda farmers could have grown.

But the visitor to modern Banda would be hard pressed to find a single grain of pearl millet. Until I started talking with people and looking at archaeobotanical samples, I had no idea that pearl millet was until recently the mainstay of people's diets, or that it remained the most ritually potent food alongside yams. Yams are the well-known centerpiece of the Yam Festival, which is celebrated throughout forest-savanna and forested parts of Ghana, most notably in Asante (chapter 3; Bowdich [1819] 1873), and is frequently commented on by outsiders. The Yam Festival marks the harvest of yams in August or September. It is a centralized affair, with people contributing their share of yams to a larger feast. In theory, no one is permitted to eat or harvest yams until the festival takes place. The chiefs use some of these yams to make offerings to ancestral stools out of the sight of most Banda villagers.

But if yams are the food celebrated and controlled by authorities, it is pearl millet that forms the focus of household-scale ritual. It is fed to household shrines and more powerful regional shrines like the ancestral baobab Wurache near Kuulo Kataa. It is the meal prepared by doting mothers to mark their daughter's transition to womanhood and marriage. Pearl millet performs essential ritual and social functions throughout the year, perhaps because in times past it was the food that fed people throughout the year.

If pearl millet is so important, why was it abandoned? In my interviews, some people blamed the prohibitions around its use and harvesting, most of which were unique to pearl millet. It is said that if people fought while planting or harvesting or processing millet, one of them would die. Some elderly people said the youth today were too quick to anger, so much so that cultivating millet would have been injurious to their health. Pearl millet processing—the process of separating the grain from the husk—was subject to a special rule, that anyone who took part in the processing should receive a share. Many people argued that this is unfair, and that under this rule, even the fetus in a mother's womb would receive its own share. A functionalist interpretation of these taboos suggest they would have acted to maintain the peace and redistribute food to those who helped in processing. These were social strategies that may have played an important leveling role in the past, especially in times of food shortage. Of course, there may also be a more mundane explanation: processing millet is very uncomfortable, as the chaff bits make one incredibly itchy, and people must coat themselves in ash to avoid this. Given the discomfort involved, people may have simply needed a special incentive to participate. Whatever the ultimate purpose of these stringent rules, in a modern setting they may have decreased the farmer's harvest enough to make millet production unprofitable. The smaller amounts available to sell would have been exacerbated by the low price fetched by pearl millet at market. Most farmers are quite savvy about cultivating crops they can both eat and sell, and pearl millet may simply not have produced enough of a profit margin. With harvests also reduced due to changing rainfall regimes, pearl millet was simply too expensive to grow.

TABLE 2 "Olden Times" Dishes of the Banda Region

Pearl millet-based dishes

Fuura	Add some water to whole pearl millet and allow to ferment for several days. A leaf called *nyaadidinge* may be added on the last day to allow it to ferment too. Grind millet and spices, sieve, and mix with water again to form balls. Boil. May pound hot balls again and add spices. Form new balls and roll in ground millet to serve. Smash and enjoy with pepper or milk. See figure 10.
Koko (all millet)	Thin porridge
Kotro papa	Grind pearl millet (usually on a grinding stone), add a sprinkle of water and mold into balls. Boil balls. Once cooked, serve by breaking open balls and adding stew of shea oil, onion leaves (*gabu*), and pepper. Often eaten in the morning. See figure 9.
Sisa/sesa	Grind pearl millet (or maize, but requires pre-frying, grinding, and subsequent sieving) on a stone. Add water during or after grinding. Pepper, salt, or sugar may be sprinkled on for additional flavor.
Tuo zafi (all millet)	See figure 7.

Cassava-based dishes

Klakro	Grind fresh cassava and squish water out (effectively detoxifying bitter varieties). Add pepper and salt to mixture and form into small pieces. Fry in oil like a donut.
Wehan	Grind *dawadawa* beans into flour and mix with cassava flour (or "chaff," scraps from sieve unusable for TZ). Mix with water, and put in pot to steam into cake. Add sauce of shea oil, pepper, garden eggs, and salt. See figure 8.

Maize-based dishes

Chobi or *brempo*	Cut fresh maize off the cob. Add groundnuts and *flewe*, a wild leaf. Cook and serve with oil and pepper
Dibudibu	Pound maize, sieve, and form into balls. Boil balls, and add sauce of onion leaves (*gabu*), pepper, oil, salt to serve.
Kwasidaman	Pound maize and sieve. Use the residue from the sieve, add water to it, and cook. Made before the grinding mill came; now impossible because you do not get the "chaff" from sieving.
Nam	During drought, people with money would buy maize and grind. They would add maize flour to water and cook until hard. Pepper, salt, and onion were added to serve.
Wenjor (all-maize TZ)	Grind maize and prepare a special *tuo zafi*. Cut and cool and save. When ready to eat mash like *kenkey* with water and add honey.
YɔɔrɔdɔOkono or *chapila*	Fresh maize is cut off the ear and pounded with salt and pepper. The mixture is folded into maize husks, which are tied shut and cooked in boiling water. See figure 11.

Other dishes

TABLE 2 *(Continued)*

Local Name (Nafaanra)	Preparation and Ingredients
Dankatere	Multiple fruits can be prepared this way. Pick ripe fruits and pound if needed to release flesh from seeds. Put into basket and add water, and let sit overnight. Run through a sieve to remove chunks. Cook over fire with quicklime and cool. Serve as beverage, especially to children.
Faaro	Prepare in the same way as *kotro papa*, but use bean flour and mold into a special leaf for boiling. See figure 12.
Gora	Pound and cook bean leaves (*susuwere*) as for leaf soup. Pound *fnumu* (seeds of *Lagenaria siceraria*, bottle gourd) and cook separately. Pound cooked leaves again and add cooked *fnumu* to mixture. Add oil, pepper, salt, and onion to serve.
Seed cakes for soup	Use *kondo* (cotton), *sisre* (*dawadawa*), or *kombotoo* (kapok] seeds and grind. Allow to ferment and form into cakes. Add to soups for extra flavor and nutrients (also see Ham 2017).
Wasawasa	Dry the heads of yam, which are usually very hard and thus rejected for *fufu* preparation. Grind into a flour and mix with water, and cook in the same way as *wehan*. Add shea oil, salt, and pepper to serve.

As an ingredient, pearl millet is not only much more nutritious that maize or rice, but it can be prepared into a wide array of foods. I offer a detailed list of these dishes, as well as a host of others made from various ingredients, and how they are prepared, in table 2. For the most part, these dishes are no longer made in the area today, and knowledge of how to make them is also disappearing. In 2009, most women in their twenties had never seen them made; women in their thirties and forties had seen their mothers make them but did not cook them themselves. Yet when these dishes were cooked for the Olden Times Food Fair described in chapter 6, people enjoyed their tastes and textures, which are quite distinct from common fare.

Describing three of these dishes, from the easiest to the most ornate to prepare, helps bring the versatility of millet to life, but also shows a very different range of culinary preparations than the everyday fare of TZ and *fufu* described above. The first is *sesa* or *sisa*, which is simply millet ground with water, though sugar can be added as well, often on a grinding stone. The Nafaanra word is also used to describe clay prepared in the same way for pottery making (see Logan and Cruz 2014 for an exploration of how these preparation techniques cross over between craft and culinary practice). This simple dish is one of the quickest to prepare, and was often provided to men going off to hunt or farm or immediately on their return. In one woman's words, it was the "olden-times fast *kenkey*," referring to the food often purchased on the go as a quick snack or meal. The food appears

FIGURE 9. *Kotro papa*: boiled pearl millet balls tossed in shea oil, onion leaves, and pepper (see table 2).

FIGURE 10. *Fuura*: heavily spiced, fermented pearl millet dough ready to be fashioned into balls and rolled in flour (see table 2).

to have a long history; the same preparation was described as fare of Asante soldiers in the early nineteenth century (chapter 3). Consuming ground flour on its own was strategic as well as nutritious; since the flour was consumed raw, there was no need to build a fire that would risk revealing the soldier's location. *Sesa* can also be prepared from maize, though maize must be fried first, making this a less efficient preparation than for millet.

The second and most commonly mentioned dish made of millet is *kotro papa* (figure 9). To prepare it, a woman grinds pearl millet on a stone, adds water, and molds the mixture into balls. The balls are boiled in water, then removed and pounded lightly in the mortar. *Kotro papa* can be dressed with pepper, *gabu* (dried onion leaves), shea butter, and salt or can be eaten as is, or stew may be served alongside it. In the past, this meal was often served in the morning. There is really no similar dish made today, though one woman did describe how a government food aid called Tom Brown (a mix of bean and grain flour) was distributed in the 1980s and made into a similar dish.

The final dish is *fuura* (figure 10), which is much more time- and skill-intensive to produce. Some considered *fuura* to be chief's food, while others described it as a prepare-ahead afternoon meal. The flavor and spice combination was also unlike any of the other dishes I encountered at Banda's Olden Times Food Fair, or indeed any I had heard described in my many conversations on the topic. *Fuura* is made of pounded or ground millet that is mixed with water, molded into balls, and left to ferment. The balls are boiled, removed from the heat, and pounded again to give them an extra-fine texture, and spices are added. The dough is then formed again into balls which are rolled in finely ground millet flour. They may be served alone or with milk. The taste of *fuura* is unlike anything I had experienced in Banda or elsewhere in West Africa. It is spicy, with a distinctive clove-like bite and tangy fermented taste, and with a very fine texture.

The current status of these three dishes, as well as the many others listed in table 2, betrays one of the most common processes in food globalization: loss and the subsequent narrowing of culinary repertoires. Foodways are always changing and such losses may well be inevitable, but in chapter 6 I ask whether some of these foods may be worth saving, and what purposes they can serve.

"SELL YOUR SHE GOAT" TIME OF YEAR

The loss of pearl millet and of the knowledge needed to cultivate and cook it raises concerns about Banda's ability to cope with climate change in the future. But on the other hand, the production of pearl millet seems untenable given current environmental, economic, and cultural constraints. At the end of the day, it is unlikely that a single crop will impede people's ability to survive, because of the tremendous variety of strategies for coping with uncertain conditions. Food security is often but not always at the heart of the tensions between tradition and modernity that I have discussed in this chapter. In this section, I attempt to bring an experience of food insecurity to light in order to underscore its complicated relationship with modernity.

Food security was a difficult topic to discuss at first. Among my first interviews in 2009 was one with a kindly older woman in Bui, a village adjacent to the archaeological site of Bui Kataa, where we had excavated the previous summer. Archaeologists often work near modern settlements, but the inhabitants of those places are rarely part of archaeological narratives, even if their lifeways often play an uncredited role in our interpretations of the past. I was fortunate that she was among my first interviewees, for I had a tremendous amount to learn and required a patient guide.

Abena Yeli, as I will call her here, was the first to teach me not to ask directly about food security. As I learned much later, admitting to food insecurity was tantamount, for some in Banda, to admitting failure, because the inability to feed oneself was often blamed on oneself. This belief is a tell-tale sign of the embeddedness and local impacts of the mantras that inform development work, as well as of the pull-yourself/your-country-up-by-your-bootstraps nationalist discourse common in Ghana. Instead, I was to ask about the changing seasons, because that is how food was best understood. Food insecurity was something better observed than discussed, better accessed through talking about contributing factors, like rainfall or short cash reserves, rather than asking about the abstract category it has come to occupy in Western discourse. The lived experience of food insecurity is not an abstraction or a wicked sustainability challenge or any of the other ways in which it is anachronized in academic and development circles. Hunger is simply experienced, dealt with, and avoided when possible.

There are a number of well-ingrained social strategies for lessening hunger's bite. Food sharing still exists in Banda, albeit in more constrained ways than in the past, and this should be regarded as an accomplishment given that food sharing has disappeared elsewhere under similar constraints (Mandala 2005). There are strict rules about sharing with kin, especially the elderly, but depending on their income level, people also regularly share with other household members and non-kin. One of the first tactics people pursue when they run short of food or money is to borrow from kin and friends. Another common way to get by is to eat foods from the bush, as described above. People also stretch out meager food supplies by eating less. They might sell the crops that fetch higher prices at market and buy cheaper alternatives like cassava. In more extreme food shortages, people might consume cassava before they had a chance to detoxify it, or skip the starch altogether and drink only soup.

In subsequent days I returned to cook with Abena Yeli. I watched as she cooked the meat I brought as a gift and squirreled the cooked meat away for future use. I was intrigued by the odd grey-green color of the *tuo zafi* she was making. At the time I thought of *tuo zafi* as a predictable dish of beaten grain porridge, but starting with Abena Yeli I came to realize the flexible nature of the food. Abena was making her food out of cassava that had been first dried and ground, as is most cassava used for TZ is. But hers had been dried in the wet season, when conditions were simply too wet for proper drying, so that mold encased the roots. I was visiting Abena during the height of the hungry season in late July, before corn or yams were ready to harvest, during the "sell your she-goat" (*Mu sikalo pre*) time of year. And cassava, she told me, is always there for you.

Only when I moved to other villages did I realize what an outlier Bui village was in terms of food preparation and food security. Part of this was the time of year—late July as the hungry season was at its worst. But the other reason was that Bui was about to be relocated in advance of flooding for the construction of the Bui hydroelectric dam. For years, government officials had told Bui villagers not to grow more than they needed, not to focus on cash crops or those which require many years to reach maturity like teak and cashew. Their fields would soon be flooded, so there was no point. The same was true of house construction. People had delayed building or repairing their own houses since they would soon be destroyed anyway.

When I visited Bui in 2009, people were willing to make these sacrifices for the good of the nation. It was their duty as Ghanaians. By 2011, when the construction of the dam was well underway but the imminent, promised move had still not happened, people were fed up. They had been living on a razor's edge of poverty and impermanence for years. Not only that, but the benefits already accruing to their neighbors, like the installation of electricity, were denied them. There were conflicting accounts about who was to blame, but people were ready to move and start their lives anew. They were the other side of modernity: the side not included. As development rolled across the Banda landscape, it was clear that its advance was patchy, unequal, and, for many, delayed (see Ferguson 1990). These hungry underbellies of modernity are often forgotten in the West's self-congratulatory stories of progress, but yet they are also modern. Adding deeper histories of remotely global places like Banda helps reveal those alternate modernities.

The advance of so-called modernity did not solve Bui's problems, and in many regards it created new ones. By the time I returned in 2014, people had been living in their new houses in the recently built Bui City for some time. I could smell fresh concrete and see the crisp lines of unchipped paint on their new, almost antiseptic, cement bunker houses. The landscape felt barren. There were no shade or fruit trees, which one finds in most villages. Those could not be moved; they had to be reestablished. But many of the town's ancestors had been moved to a burial plot not far from the houses, thanks to the involvement of archaeologists from University of Ghana. Still, even to an outsider like myself the town felt deflated. Life in the absence of a meaningful cultural landscape was quite different, as captured in Devin Tepleski's documentary *Mango Driftwood* (2010).

The new Bui City was something of a political experiment as well. The town was an amalgamation of Bui village and three other villages. The designation "city" was perhaps intended to capture this diversity, but in practice the villages operated as separately as possible. In addition to the quarters occupied by each village, another cluster of houses stood off to the side. These houses were larger, cordoned off by fences, and had their own water tanks, signaling the availability of running water inside. They were clearly better than the concrete bunkers occupied by Bui peoples. Bui villagers were bitter. These houses were not for Africans at all, but for foreigners coming to work on the dam. Tensions between the diverse occupants at Bui were high, and did not show any prospect of abating.

Development is not a panacea. Often, it exacerbates old inequalities—between chiefs and constituents, haves and have nots, women and men—and can also act to create new disparities (Carr 2008; Cliggett 2005). As Bui's example illustrates, food in particular is often strongly impacted by development schemes and modernization. From the long-term perspective offered in this book, we can see that developments like the Bui dam are sedimented on top of earlier shifts like market integration, building on their imperfections, deepening the crevasses of inequality. We must understand those histories in order to work towards a better future, a subject I continue in the final chapter.

HUNGRY UNDERBELLIES OF MODERNITY

In this chapter, I have covered only a small number of the many tensions that inhabit remotely global foodways in Banda. I have tried to coax to the surface the decisions people make on a daily basis in order to feed their families in a manner that subverts the scarcities they face. This subversion is two-fold, as it involves satisfying nutritional needs despite declining food availability, while also lessening scarcity's bite by preparing and consuming desired and culturally valued foods. Navigating this tension means negotiating between the two definitions of scarcity discussed in chapters 1 and 2, of measurable, quantifiable need and the perception of need. As it turns out, eating while remotely global involves balancing both.

In some cases, people have avoided both scarcities by adhering to tried-and-true traditions, like wild leaf soup and *tuo zafi*. This adherence may at times be a conscious choice, at others an unintentional one grounded in bodily reproduction, and at still others a function of less choice brought on by poverty. Likewise the appropriation of foreign foods and things falls on a similar spectrum of choice and lack of choice, especially in this remotely global place where funds do not always permit access to the most desirable foods, like rice or fresh meat. Yet people continue to choose what is best for them based on their circumstances at the time; such appropriations are not disempowering capitulations to outside pressures, but active strategies that build rapidly evolving local tastes.

Balancing real scarcities by avoiding the perception of hardship is a dialectic, always in flux, and appreciating that constant movement means we cannot regard African food practices as timeless and unchanging no matter how traditional they may appear in the eyes of outsiders. As Piot (1999) has argued, life in an African village is just as modern as in London. Africans have been enmeshed in the very same global systems as British city dwellers for centuries, but the power dynamics have been unequal for a very long time. Out of these inequalities two modernities were born: the well-fed and the hungry underbelly. In these hungry underbellies of modernity, people regularly deal with seasonal food insecurity. But at the same time, they hunger to be part of the other modernity, the well-fed one. Aspiring to eat like urbanites, with their ready access to mass-produced convenience foods, helps the youth in Banda feel like they are part of the modern world. In the concluding chapter, I consider how these modernities might be reconciled.

6

Eating and Remembering Past
Cultural Achievements

People in poor countries are slow to develop because their cultures do not develop as fast as the cultures of developed countries . . . . The cultures of the poor countries do not change fast because they do not assimilate and use new ideas and new ways from other cultures that will help to improve their way of life as fast as possible.

—Text from 2009 GHANAIAN SOCIAL STUDIES TEXTBOOK, DISCUSSED BY ROCK (2018)

A generation ago, Terence Ranger (1976) described attempts to make African history usable to contemporary African peoples. He focused on the largely political projects of African leaders of early independence, who showcased past achievements to instill pride in African history and identity as part of postcolonial African nation-building. Bassey Andah made a similar pitch to archaeologists in 1995, although in more recent decades what is now known as "usable pasts" archaeology has trended towards more scientific applications, like potential uses of ancient agricultural technologies to improve production in the present day (see Logan et al. 2019). Yet whatever their lofty ambitions, rosy valorizations of the past can be at odds with the people they are supposed to help, which forms the subject of this concluding chapter.[1]

Younger generations often take issue with initiatives designed to promote the past for use in the present. As Ranger (1976, 22) put it, "the young radicals object that the poor and hungry cannot eat past cultural achievements." The "young radicals" of Ranger's day posed precisely the same question that many of my interlocutors did: "Of what use is this project to us?" Most people meant this in practical terms—how does asking about past food changes help my family eat today? But the younger generations also meant it in ideological terms. Tradition is a dirty word to many of the youth I have spoken with in Banda. It is associated with backwardness and plays the foil to "modern" developments in the rest of the world. To "keep with the old," as many of my young Banda interlocutors put it, is to lack vision for the future. The sentiment is not unique to Banda, as Piot demonstrates in neighboring Togo (1999, 2010) and Ferguson (2006) details in Lesotho. It is reinforced and (re)created by sources near and far, including in Ghana itself, as the excerpt above demonstrates.

The automatic response of many archaeologists and anthropologists to this kind of mischaracterization is to defend the past. Here caution is warranted. Not all pasts need resurrecting, and sometimes people prefer that the past stay in the past. When Ferguson (2006, 19), for example, extolled the virtues of traditional earthen architecture over what he saw as poor copies of the colonizer's building styles in Lesotho, he was told that those "copies" were claims to "a direction we would like to move in." Continuing to build earthen houses was not a choice, but symptomatic of a lack of options. I encountered a similar dynamic in Banda during a vernacular architecture study I pursued in 2011. It reminded me that my own interests and priorities were often very different from those of the Banda community.

With this proviso in mind, this final chapter aims to provoke questions rather than prescribe solutions. Can people eat past "achievements," and if so, how? And how should the past be remembered, and to what ends? I prod the first question by considering the interplay of the practical and ideological barriers to eating "olden times foods" in Banda. In the second and final section, I trouble the status quo with radical reimaginings of food and agriculture based on the deep history of food security covered in this book.

"FOR ALL TO SEE": THE STIGMAS
OF HERITAGE FOODS

Early one July morning in 2014, about thirty women assembled outside the Banda Cultural Centre.[2] Within short order, they set up multiple kitchens, including three-stone hearths, pots, pans, plenty of food, and wooden mortars. Smoke soon filled the air as did the rhythmic thump-da-thump of large pestles crushing grain in the mortars. People of all ages came to observe the cooking on display. Queen Mother Akosua Kepefu and the women elders had insisted upon leaving the convenience of their own kitchens in order to show the younger generation how to make "olden times foods" in a public and socially sanctioned arena. The choice to be on display, "for all to see," as the queen mother put it, was an important one. Public visibility not only helped promote these little known dishes, but also began the process of destigmatizing olden times foods and the women who prepared them (figures 11, 12, 13, see table 2).

FIGURE 11. *YɔɔrɔdɔkonO* (*chapila*): fresh maize dumplings steamed in maize husks (see table 2).

FIGURE 12. *Faaro*: bean dumplings steamed in leaves (see table 2).

FIGURE 13. Making food for the Olden Times Food Fair.

I had encountered some of the challenges that women faced over the course of many interviews as well as in my repeated visits to the area. Most of the old women with whom I spoke in 2009 were among the poorest in their villages. Some were in ill health and passed away during my few years of absence. Yet they were repositories of knowledge and were delighted to be asked about it, since many of their experiences had not, by local standards, been considered part of the history that gets to be told.

In 2014, myself, Enoch Mensah, Zonke Guddah, and several community leaders organized a joint Heritage Day and Olden Times Food Fair.[3] My goal was to provoke conversations about the use of the past in the present, particularly in the context of ever-widening generational divides. Beginning in 2007, the construction of Bui hydroelectric dam in the region had caused an almost fast-forward version of development in the area. New roads, electricity, and cell phone towers had snaked in, as did cheap wage labor at the dam site itself. During my fieldwork in 2009, frequent altercations broke out over the low pay and long hours there. The youth benefited from increased access to cash, but this only increased the gulf between them and their grandparents. The Heritage Day event, we hoped, might bridge this gap by generating conversation about what can be learned from the past. I approached the event with the hope of opening lines of communication between old and young, traditional and modern, rather than with a specific prescriptive goal.

As the Heritage Day event began, the early morning food preparation out-side of the cultural center served to draw a crowd of curious onlookers with new smells—and the promise of free food. While the set-up crew played music on the enormous speakers common at Ghanaian events, the women elders provided a sensory sound- and smell-track of food preparation that lured the youth to the Olden Times Food Fair (figure 13).

My choice to emphasize food was deliberate. Initially I had thought foodways were a relatively uncontroversial form of heritage that would stimulate conversa-tion and cohesion. I soon learned that food, like all forms of heritage, is contested and negotiated (Di Giovine and Brulotte 2014). Food is grown or prepared by people, and the social values assigned to individual cooks and farmers are often extended to the food itself. These social imbrications give food its value. Heritage foods often carried negative social stigmas, a value that was very much at odds with how I, as an outsider, perceived them. This divergence of views is worth exploring, because it provokes important questions about the future of food in African contexts.

In my earlier interviews, in 2009, I remember being struck by how often my female interlocutors brought up female circumcision during interviews on food. At first an odd pairing, the connection makes sense in light of my goal at the time, which was to contextualize shifts in foodways with major changes in wom-en's lives over the last generation or two. Cessation of female nubility rites, which included circumcision as well as a constellation of other practices that marked a girl's transition into womanhood, was the most often reported significant change in women's lives. These practices formally ceased in the 1990s, because of both a national law that outlawed female circumcision and an active campaign on the part of the rapidly growing local Christian community to cease the practice. Female elders had been in charge of nubility rites, and the power and respect their positions demanded diminished when these practices ended. Not only did elderly women lose this source of authority over younger men and women, but many became increasingly stigmatized as Christianity gained traction.

The female elders themselves were a big part of why heritage foods were stig-matized. The nubility rites controversy illustrates how older women in authority positions have often been on the opposing side of Christianity, now a powerful force in Banda and throughout Ghana. This opposition came up in many conver-sations with my interlocutors in 2009, and seems to have grown more acute in the years since then. Many of these women feel they are no longer treated with respect, because they do not attend church and are associated with the "old ways."

During the event and even afterward, we overheard numerous conversations about whether or not the olden times foods presented were safe to eat. People were scared that the elderly women who made them—women who were themselves stigmatized for their spiritual choices—had poisoned or cursed the foods. At first, people went so far as to avoid some of the foods for fear that eating them would make them ill. After some discussion, most decided that since these foods were prepared out in public view in front of the cultural center, the suspect women would not have been able to curse or poison the foods. In other words, the queen mother's insistence that these foods be prepared "for all to see" was an insurance policy against later accusations of illness as well as an attempt to increase the likelihood that people would feel safe eating the food. The foods, in other words, were seen by some as imbued with the stigma of their makers.

These kinds of accusations against women in particular have long histories in Banda; archival records, for example, document accusations of poisoning against Abena of Banda in the early twentieth century (chapter 4). Women's control over the intimate domain of consumable substances is one source of their power, but also a channel through which accusations can easily flow (Douglas [1966] 2003; Lyons 2014). It was these stigmatized women who were my core interviewees and later collaborators in the Olden Times Food Fair. They valued history more than many, thus were ready-made interlocutors and collaborators. My focus on food rather than on topics that are the more traditional domain of history (e.g., the lineages of chiefs) meant that at least some of their knowledge got to matter. In the design of the associated Heritage Day event posters, I consciously attempted to combat the stigma against them by promoting women's histories in particular. But like all forms of prejudice, stigma has a certain staying power.

The stigma against olden times foods also relates to the association of some heritage foods with scarcity. During both severe and seasonal food crises, one of the most common coping mechanisms is the increased consumption of leaves from both crop plants and the bush. These are most often cooked into a kind of sauce or soup, and, as described in chapter 5, add considerable diversity and nutritional value to otherwise starch-heavy local diets. *Gora*, a mixture of bean leaves and *fnumu* (squash seeds), is one such way of making food stretch farther. The queen mother remarked that *gora* was often used as a meat replacement in times when meat could not be purchased or hunted. In focus groups that followed the Heritage Day event, both old and young people remarked that foods like *gora*—those consumed out of necessity—would not grow in popularity.[4] This is not because their taste was bad, but because they are associated with poverty. No one would purchase such foods from public street vendors because they were poor foods.

Continued preparation of foods like *gora* can be interpreted as motived by a desire to resurrect or uphold tradition, but we must be attuned to instances where, as Ferguson (2006, 21) puts it, a "traditional African way of life is simply a polite name for poverty." While poverty is not unique to elderly women, they are often among the poorest in most villages. Food-sharing norms in Banda are that a daughter-in-law provide food for her mother-in-law, a more still honored in much of the Banda area. But when times get tight, as Cliggett (2005) describes in Zambia, it is often the elderly women who suffer first and the most. It's no wonder they maintain their knowledge of necessity foods. The challenge is how to impart this wisdom to the younger generations who may someday need it themselves, but would not be caught dead eating or preparing poor foods (even if they are consumed out of sight). These necessity foods are associated with the recent past, back in what many youth termed the "days of ignorance."

A straightforward appeal to tradition does not impart heritage foods with value in Banda; this distinguishes African edible heritage from its counterparts in places like France or Italy. People practicing the most "traditional" lifestyles are often among the poorest. They have a much more narrow set of choices than those who are more affluent. Tradition, in this way, becomes associated with backwardness and scarcity. This situation is only compounded by external views of Africa as a place lacking in both resources and expertise. For these reasons, necessity foods may not be the way to promote African edible heritage. The youth will not eat them if they have any choice. Knowledge of such strategies can and should be maintained, but this will likely be done in secrecy, since people hide their struggles with meeting subsistence needs.

Many of our subsequent focus group discussions deliberated about what needed to happen to make olden times foods sell. Monetization of these foods is critical, and not only because the women who know how to make them are usually in dire need of cash income. The other reason is that most people encounter new and different foods when they patronize street vendors. In the context of interviews on food change that I conducted several years before the Olden Times Food Fair, I came to appreciate that while consistency and labor were valued in the home kitchen, novelty was usually experienced in the context of purchasing breakfast or lunch when out of the house for work or school. That our focus groups narrowed in on street vendors acknowledges this opportunity for the introduction of new (old) foods.

However the youth in these groups were quick to point out that olden times foods would have to be modified in certain ways for them to be desirable and sell. One simple way was to promote those foods that do not carry the stigma of poverty, as *gora* does. One such food is *fuura* (chapter 5), which was a luxury food of the chiefs in the past. It's no wonder why: *fuura* is incredibly time-consuming to prepare, as it's twice pounded, twice cooked, and heavily spiced. The resulting product is of a very fine, desirable texture as well. The high labor input required would, we hoped, mean that *fuura* would fetch a higher price, but even this would depends on how it was valued.

One other option was to promote quickly made, tasty convenience foods like *sesa*. *Sesa* is simple to prepare; it is essentially ground maize or millet mixed with a bit of sugar. This meal was widely consumed in the past by men and women on the move. Nineteenth-century accounts suggest it was the food of soldiers on long campaigns, as it did not require a fire to prepare (chapter 3). In the focus group, one woman remarked that she had sold *sesa* on occasion with great success, by mixing the flour with Nido milk powder and packaging the *sesa* in a plastic bag. Both innovations make the food more palatable because it appears and tastes more "modern." Other women suggested that some of these heritage foods be made more modern by swapping out shea butter, a cooking oil with a strong taste, for mass-produced vegetable oil, which has come to be widely used in the area. This kind of suggestion was also observed by Ham (2017) in northern Ghana as a desire for the more neutral smells and tastes of foods associated with modernity.

Some of the young men who took part in our focus group suggested that none of these foods would sell unless their look was "modernized." What they had in mind was a repackaging of sorts, so that the olden times foods would resemble industrialized food products. Why? Because then "modern people"—a group often juxtaposed against elderly women, described as "olden times people"—would buy them. This is a strategy that some women observed in other regions: repackaging of local foods made them more appealing to younger consumers. This repackaging would give these foods an aesthetic of modernity, and all of its assumed benefits, including hygiene. But I suspect there is much more at play in giving these old foods a modern look. Doing so disassociates the food from both forms of stigma I discussed. Buying packaged foods suggests people have money to do so, obscuring the roots of some of these foods in times of scarcity. Packaging also obscures the maker of the foods, suggesting an industrial-scale operation rather than an elderly woman suspected of witchcraft. Through being wrapped in plastic, olden times foods are sterilized of their association with the spiritually liminal.

The resulting aesthetic is very different from what Western consumers might expect of "local" foods. Admittedly, I have very few photographs of processed, packaged foods in Ghana precisely because they did not appeal to my aesthetic. In European and American settings, local food consumers pay top dollar for foods that come from specific farms or regions, carrying with them *terroir*, or taste of place (Trubek 2009). We place value on knowing who made our foods, and under what conditions. Our packaging reflects these values. But to the African consumers presented in this study, the values desired represent the other side of the coin. In nearby Togo, Charles Piot (1999, 2010) has documented how elements of globalization are taken up and given new meaning and context locally, not unlike how foreign objects were handled in the past. In case of Banda, modernity takes on a plastic form, enveloping stigmatized foods and rendering them edible. The aesthetic here signals edibility because it indexes how the food was made: in a modern, hygienic factory, not by your grandmother who doesn't go to church.

CHALLENGING GLOBAL DEVELOPMENT DISCOURSE
BY REMEMBERING THE PAST

In essence, Banda youth were not asking *if* we can eat past cultural achievements, but whether we should bother at all. Some agricultural development experts, like the USAID officer presented in chapter 5, might agree. Both parties often hold indigenous African foodways in low esteem. And although their motivations for holding these views differ, this convergence merits discussion. The ideas of Banda youths about tradition and the past stem not only from local history and experiences of poverty, but also from a global discourse of which they are as much a part as USAID or Monsanto. The textbook example that opened the chapter illustrates what Ghanaian youth learn in school: that all innovation comes from the outside, and that African traditions must be abandoned to move forward in the world. This view is remarkably similar to views of the USAID officer described in chapter 5. Both stem from the body of accepted wisdom about the past that I call the scarcity slot.

The negative views assigned African foodways are also closely intertwined with gender. In a recent piece in *Popula*, Hamza Moshood relays this dynamic in her account of "Our Day" celebrations in Ghanaian schools, in which pupils bring food to class.[5] Plates are dominated by rice and biscuits—foods that must be purchased—rather than local, indigenous foods, which are looked down upon. This dynamic is as deeply gendered as it is historical. Women often cook and serve food in chop bars, ubiquitous food stalls where inexpensive local fare is offered. While men consume this food, men do not work as cooks or servers at chop bars. But it is acceptable for them to prepare and serve "white man's chop." The association of indigenous, local foods with women is clear in Banda as well, and relates to ways in which women are devalued in general.

The dense interconnectivity of the discourses of Banda youth and those of global development forces suggests that any attempt to build usable pasts must be accomplished on multiple levels. Events like the Olden Times Food Fair, or any other kind of heritage-promoting initiative, must be part of a broader push to change discourse at the global level. Past cultural achievements can be used to instill pride in local and national history, but to be effective they also need to challenge global misconceptions.

Today there is arguably a greater political need to promote the qualities and capabilities of local African foods than ever before. Multinational biotechnology corporations are making a run on Africa's genetic crop resources, and their local devaluation only facilitates this knowledge theft. Climate change projections predict a dire decrease in yields if the continent's agricultural production as practiced today does not change. This has served as a call to action for various multinational biotechnology corporations, which claim that Africa has the most to gain from GMO technology. Ethnobiologists and critical development practitioners have mounted fierce critiques of these views, illustrating the importance of preserving local landraces and ethnobotanical knowledge. Yet they face an uphill battle, especially from the better-financed corporate propaganda machines, which accuse activists of withholding live-saving crops from their constituents. Activists in turn accuse seed multinationals of neocolonialism (Rock 2019).

Biotechnology is the newest iteration of transnational development machines, whose very existence depends on the scarcity slot. A worldview in which Africans lack food and the ability to do anything about it themselves because they are hope-lessly stuck in the past justifies and legitimizes international development and biotechnology in particular. But what if we consider an alternate reality where such stereotypes are replaced instead with empirically grounded narratives like the one presented in this book? What would this version of reality look like, and how would it change how both Westerners and the youth of Banda think about African food and agriculture? In what follows, I present a hypothetical imaginary to prompt critical reflection on how stereotypes loom over approaches to the con-tinent's needs. This kind of imagining reinforces precisely why history matters for resetting future possibilities.

· What if the "Africa" we heard about on the news was one that had pioneered sustainable, organic agriculture?
· What if Africans were looked to as sources of inspiration for dealing with twenty-first century climate change, given peoples' long histories of experimentation and successful adaptation to climatic variability?
· What if regional African foodways were revered in the same manner as French cuisine, and African terroir and expert chefs were lauded for their creativity and evolved methods?
· What if the superior nutritional content of West African diets was held up as the gold standard to which we all should aspire?

The thing is, none of these "what ifs" are fantastical. Each of them is grounded in empirical research. The consistent, repeated, and wholly entrenched stereotypes about the continent that I call the scarcity slot are the only reason that none of them are commonplace views in today's world.

So what would development look like in modern Africa if the "what if" story was the one that got to be told? Africans would have to be approached as experts rather than passive recipients. European interventions would have to be blamed for upsetting an otherwise sustainable system. Fingers would be pointed at the unevenness of global trade and who gets to benefit and why. In other words, accountability for Africa's food problems would be transferred from Africans and their environments to Western interventions, just as Rodney (1972) tried to do decades ago.

In this version of reality, Banda's culinary history showcases flexibility and per-sistence in foodways during some of the most severe environmental, political, and economic crises the world has ever seen. This narrative of resilience not only raises the specter of possibility for home-grown solutions, but also argues that we need to search for the root of Africa's food insecurity problems in other places. From the long-term perspective adopted in this book, it is clear that local foods and local knowledge have provided Banda the means to maintain food security even during the most severe drought on record in the last millennium. This success story holds salience for the climatic challenges that Banda farmers are currently facing, as well as for those predicted to cause grave food security challenges in the continent in coming decades.

Banda is but one point on a map. We desperately need to build *longue durée* histories of food security in other regions on the African continent, to capture the range of intersections of climate, history, and politics. In this book, I have argued for an empirical, interdisciplinary approach that builds on the strengths of history, archaeology, anthropology, and environmental studies, while also being critical of the assumptions these disciplines bring to the table. Disciplinary silos may lead to the uncritical reproduction of implicit prejudices, in part because of the limitations of each of our archives. As Africanists have long showcased, comparison of different source materials often reveals the fissures and tensions between disciplinary interpretations, and has provoked new kinds of questions about the past and its relationship to the present.

The past is an effective foil to the present because it can be used to expose and elevate African possibilities and capabilities. Yet, as the Ghanaian social studies textbook quote betrays, history is viewed as largely irrelevant in modern Africa. This "forgetting" has long been one of the central projects of imperialism, and is one of the most long-lasting and powerful impacts of colonial rule (Santos 2018). Forgetting is in many ways the engine of the scarcity slot. It is time to remember those pasts. It is time to raise the limits of the possible in order to reinvent food sovereign futures. That such futures are possible is suggested by the persistence and flexibility of rural foodways as well as by the longer-term histories accessed in the archaeological record. Even difficult pasts that attest to the structural and slow violences that have created modern-day insecurity have a role to play. These pasts show us that at least in Banda, food insecurity is a condition that was made, not one that has always been. Banda's histories also challenge us to recalibrate the association of modernity with progress. And they show us that the targets of development and biotechnology—Africans and their agricultural practices—are not the problem.

Methodology

ARCHAEOLOGICAL METHODS

The bulk of data considered in this book is archaeological, since for the majority of the period covered this is the primary source of information about food. Archaeological data are derived from excavations in the Banda region that have been ongoing since 1989. Domestic structures, kitchen contexts, midden deposits, and craft-working areas were extensively sampled at four sites (Banda 13, Ngre Kataa, Kuulo Kataa, and Makala Kataa) spanning the period from 1000 to the 1920s, with limited test excavations at many more sites as part of a regional testing program (BRP 2002; Smith 2008; Stahl 2007). Botanical, faunal, ceramic, and metal samples were systematically collected from all deposits.

All methods used in excavation of archaeological sites and processing of material culture have been previously summarized by Stahl (1999b, 2007). Middens were usually sampled by isolated 1x2m or 2x2m units (Stahl 1999b, 11), which was appropriate given the large quantity of materials that resulted. For domestic and craft-working contexts, adjacent, mostly 2x2m units, were excavated to achieve broader areal exposure of deposits. Excavation units were identified by the coordinates of their northeastern corner on an arbitrary site grid. Excavation commenced in arbitrary 10 cm levels unless there were natural or cultural boundaries present; in the case of multiple deposits or features in a single level, individual deposits were removed separately and designated differently (Area A, Zone A, etc.). Identifiable features such as pits or floors were excavated separately. All soil was sieved through a 5 mm (1/4 inch) screen (Stahl 1999b, 11–12). Pottery, bone, slag, and small finds (beads, metal, etc.) were collected and recorded separately.

MACROBOTANICAL ANALYSIS

The charred seeds, nut shells, and other materials recovered from flotation are referred to collectively by archaeologists as macroremains or macrobotanical remains. Paleoethnobotanical collection and analysis methods followed published protocols (Pearsall 2015; Piperno 2006), with modifications made for field conditions. The Banda sites are among the most intensively sampled for plant remains in the African continent. Scatter samples of 5 or 10 liters were collected from almost every level, feature, and depositional unit over the two decades of Banda Research Project excavations (1989–2009); point samples, usually of smaller volume, were collected from features of interest (e.g., burnt areas) (Stahl 1999b, 12). Interior contents of pottery vessels were also routinely collected and floated. Generally, 10 liter samples were collected in earlier phases of the project (Makala Kataa: BRP 1989, 1990), but sample sizes were reduced due to practical considerations in the mid-1990s (Makala Kataa: BRP 1994; Kuulo Kataa: BRP 1995, 2000; Banda 13, Ngre Kataa, and Bui Kataa: BRP 2008, 2009). Samples of 5 or 10 liters were the maximum feasible size given that at the end of an excavation day samples must be head-loaded to vehicles and manual flotation was the only method possible without running water (or electricity until 2007) in Banda; all flotation water was drawn from boreholes and head-loaded by women. Given these constraints, the number and size of samples collected is exceptional in West Africa. Samples were floated manually, with sediment placed in clean water in a headpan, and floating remains skimmed off using a tea strainer (approx. 0.5 mm mesh) and deposited onto finely woven cloth for drying.

Sampling of light fractions was necessary as excavations resulted in over 1,600 floated soil samples. Judgmental sampling was chosen over random sampling given the hit-and-miss nature of macrobotanical preservation (i.e., requiring fire); the diversity of excavated contexts (e.g., domestic, ritual, ironworking); and their appropriateness for answering the research questions. All samples were selected based on a detailed reading of the contexts available in field notes from each site. An attempt was made to complete a similar number of samples from each site and time period, but samples at some sites (Banda 13 and Kuulo Kataa) were remarkably homogeneous, leading to diminishing returns. Once diversity had leveled off at these sites (i.e., further samples resulted in data redundancy), I opted to sample sites with better preservation, better contexts, and more diverse assemblages. At all sites, both midden and domestic contexts were targeted. Middens were selected since they are typically richer and more diverse and allow for chronological reconstruction. Domestic contexts, though considerably more sparse, were selected in order to reconstruct daily practice over space. A limited number of samples were completed from craft and ritual areas to compare with household sites. To date, 326 flotation samples from twenty-three contexts at nine sites have been analyzed, a portion of which, from the Ngre to Late Makala phases, is reported in appendix B (see Logan 2012 for results from earlier periods). Contexts dating to the Late Ngre, Kuulo, and Early Makala phases were targeted in order to track the impact of American crops.

All light fractions were weighed, separated through nested sieves (2 mm, 1 mm, 500 μm, 250 μm), and examined under 7–40x magnification. All components (charcoal, seeds, other) were separated and recorded for the 2 mm and above fraction; the remaining size fractions were scanned for seeds only. Samples weighing more than 25 g were split using a riffle splitter; weights reported in the data tables (appendix B) are of the material analyzed only (i.e., they do not include unanalyzed portions). Heavy fractions were also collected and sorted. I made identifications by reference to collections at the University of Michigan Ethnobotany Laboratory as well as the African Archaeobotany unit at Goethe Universität in Frankfurt.

Two categories used in this study need explication. "Unidentified nonseed charred plant remains" refers to material that is clearly not wood charcoal, and may include materials of interest that could be further identified in the future (e.g., parenchyma). "Unidentifiable seed fragments" refers to seeds and seed fragments that are unlikely to be identified in the future due to extreme distortion and/or fragmentation.

Based on the generally low seed counts as well as unevenness in the data (e.g., several thousand grains in MK 6 samples versus a few dozen in NK Mound 7), I opted not to use statistical techniques for quantifying my results. Instead, I use simple presence/absence and count data. The idea was that phytolith and starch analysis would help confirm the spatial distribution of important crop and wild plants, though, as I now describe, this confirmation is still a few years off.

PHYTOLITH AND STARCH GRAIN SAMPLING AND PROCESSING

Sampling for both phytoliths and starch grains differed at various stages of the project. There are no soil samples available for the earliest excavations (in 1989 and 1990), though some grinding stones were archived unwashed—at this point in time, phytolith studies were only just emerging. Soil samples were taken more regularly in 1994 and 1995 at Makala Kataa Station 6 and Kuulo Kataa, often in the form of column samples from select walls. Soil and unwashed grinding stones were also collected from Makala Kataa Station 6 in 1994. After I joined the project in 2008, more extensive sampling for microbotanical remains was undertaken (i.e., at Bui Kataa, Ngre Kataa, and Banda 13). A small soil sample was collected from each 5 liter flotation sample in an attempt to get a scatter sample of each excavation unit. This was supplemented by point samples of interesting features (e.g., floors, hearths). Further details about the methods used to process soil samples, as well as the limitations and challenges of analyzing soils from archaeological sites in Africa can be found in Logan (2012).

The phytolith data I discuss in this book derive from artifact surfaces, with the goal of obtaining a basic understanding of food processing. Although only a fraction are reported here, 64 artifacts were sampled for phytoliths and starch using the methods outlined in Pearsall, Chandler-Ezell, and Zeidler (2004), with modifications made for field conditions. These included the creation of a clean zone in the laboratory, use of new disposable materials for each artifact (toothbrushes, plastic bags, etc.), and use of bottled water for sampling (distilled water was not available). Sampling involved collection of three sediments from unwashed artifacts (dry brush, wet brush, and sonicated fraction). All artifacts sampled were photographed. This kind of intensive sampling was undertaken for the sites of Ngre Kataa, Bui Kataa, and Banda 13. Due to differential sampling, I focused analysis on the two excavation areas with the most comparable samples: Makala Kataa Station 6 and Ngre Kataa Mound 7.

Phytolith extraction and scanning methods follow those used in the University of Missouri Paleoethnobotany Laboratory (Pearsall 2015), modified for extraction of calcium compounds, such as faunal spherulites, which are found in animal dung and are composed of calcium carbonate (Coil et al. 2003; see Logan 2006, 30–38 for procedure detail). Phytolith samples were examined at 400x magnification using a Leica DME microscope.

Unfortunately, given the infancy of phytolith analysis in Africa (especially West Africa), we do not yet have strong diagnostic indicators that would distinguish sorghum, pearl millet, maize, and other likely crops (with the exception of banana). I describe the progress that has been made in my dissertation (Logan 2012) and a recent article (Ball et al. 2016), but my general approach here is to use indicator methods. Indicators may or may not be diagnostics, since phytolith production in the local flora is not well-enough defined to evaluate possible redundancies. Instead, I assign the identification a weak, moderate, or strong probability that it represents the particular species. These descriptors indicate the strength or likelihood of the identification. *Strong* identification means that the phytolith shape has also been found to be unique across other world floras or that within any given sample, there are multiple and recurrent indicators of that taxa (especially for the grasses). Identifications labeled *moderate* have a good probability that they represent the species of interest, assessed based on their diagnostic level in at least one other tropical flora, apparent uniqueness of the form based on global phytolith literature and/or the Banda flora studied, or common presence in a given sample. *Weak* probability identifications mean that forms observed in a taxon of interest are also observed in the archaeological sample, but similar forms may be observed in closely related taxa, or closely related taxa have not been studied and may produce similar forms. Large variant 1 crosses that are diagnostic of maize leaf in the Americas offer a good example: while concentrations of these might weakly or even moderately suggest the presence of maize leaf in a sample, isolated occurrences could represent maize, or a number of other African wild grasses that produce this form.

Although phytolith production patterns in maize are well documented, only a handful of studies have documented phytolith production patterns in sorghum (Logan 2012; Radomski and Neumann 2011; Madella, Lancelotti, and García-Granero 2016) and pearl millet (Logan 2012; Madella, Lancelotti, and García-Granero 2016). What is clear is that it is possible to separate the domesticated grasses from each other; millet produces small bilobates and crosses, while maize and sorghum do not. Sorghum produces several complex short cells which are generally taller and not elongate, including the potentially diagnostic saddle-like rondel. Maize produces more elongate and squat rondels than sorghum, in addition to three forms which may be diagnostic.

The real challenge is distinguishing these taxa from wild grasses in the archaeological record. My approach here is to use a series of ratios to quantify strong, moderate, and weak probability identifications of sorghum, maize, and millet. The method should be considered tentative until independent means are used to test their accuracy. The best check I have at present is whether or not charred grains of the same taxon occur in similar contexts, but since the objective in using phytoliths is to find processing activities that did not involve fire or whole grains, this is problematic. The identification potential is highest for sorghum, followed by maize; millet remains a challenge. However, it does allow me to tentatively evaluate whether sorghum and maize are more ubiquitous than the macrobotanical record suggests.

ETHNOARCHAEOLOGICAL AND ETHNOGRAPHIC METHODS

Ethnoarchaeological methods were built around my central research questions regarding continuity and change in food practices. This involved documenting the variability in specific foods, materials, and techniques over space and between generations. I attempted, so far as possible, to avoid asking questions that would introduce assumptions of timelessness into the data. In other words, the objective of this research was not to collect baseline quantitative data for quantities such as crop yields, because such measures are not comparable or obtainable from the archaeological record, and they do not take into account the impact of changing governmental and economic inputs into agriculture (such as fertilizer), not to mention shifts in labor and gender dynamics. This is not to belittle the many excellent case studies of this type that have already been accomplished; but this is not one of them. Although I have spent a total of about eleven months in Banda, this comprised several nonconsecutive visits between 2008 and 2014, some of which were spent engaged in archaeological research. My time was concentrated in the wet season (April through October) and early dry season (October and November). Ideally, I would have stayed over the dry season, where I would have seen fields prepared for planting, and eaten the leaves and vegetables dried in the wet season for later use. However, I feel that I have an adequate idea of these activities from interviews and from previous fieldwork during the dry season (January through March) in neighboring Togo and more distant Senegal. Further, I observed and have interview data about the two primary food seasons.

My ethnoarchaeological research design was shaped by David and Kramer (2001), and was based on my experience in Sudan with Catherine D'Andrea in 2005. Prior to fieldwork, my methodology and sample questionnaires were approved by the University of Michigan's IRB, which issued an exemption, because the information to be collected was deemed not sensitive. As part of the interview process, name, age, and area of origin information were collected to aid in possible future longitudinal studies, but information was coded and/or all identifiers were removed for reporting purposes out of an abundance of caution. Names are revealed in the text where information is not sensitive, and changed in cases where the information is potentially sensitive.

My initial ethnoarchaeological study focused on food change, and took place from July to November 2009 in the period as the wet season transitioned to the dry one and food shifted from TZ to yam *fufu*. I conducted 120 interviews primarily with women spread across six villages in the Banda region, including one on the west side of the hills (Dorbour) also studied by Cruz (2003). Interviews on food change were semi-structured, initially using a questionnaire to guide the interview, but were tailored to highlight the knowledge of each interviewee. All interviews were conducted by me and translated by Enoch Mensah, a long-time Banda Research Project assistant who is fluent in the local language, Nafaanra, as well as Twi and English and is well known in the community. Generally, interviews focused on one person at a time, but some group interviews were also conducted. Men were often interviewed in groups, as they wished to give me the "official" story. Middle-aged and elderly women (forty years and older) were the focus group for most interviews, but in each village a small number of men and younger women were also interviewed. Informants were generally interviewed at their home, and selected mostly at random or on occasion based on recommendations from other informants. Interviews generally lasted around one hour. All photographs were taken with consent.

The six villages were selected based primarily on their proximity to archaeological sites, not so that a direct analogy could be constructed per se, but because residents of these towns were already familiar with the work of the Banda Research Project. Villages interviewed included Bui (near Bui Kataa), Dumpofie/Kuulo (near Kuulo Kataa), Ngre/Nyiire (near Ngre Kataa), Makala (near Makala Kataa), Banda-Ahenkro (where recent middens and a house were sampled for plant remains), and Dorbour, the latter so I could compare ethnoarchaeological observations of food to Cruz's (2003) mid-1990s examination of pottery. The good relations that the principal investigator Ann Stahl has established in the region over the last thirty years, and in these villages in particular, made it possible for me to visit several villages and quickly obtain blessings and cooperation. As is custom in this part of Ghana, before any research began, the paramount chief and elders in Banda-Ahenkro proper were consulted. Their permission secured, I then met with the chief and elders of the individual villages in which I worked to obtain appropriate permissions as well as advice on who best to talk to. Only after this did I visit individual informants and begin interviews. At all steps in this process, the good working relationships and long-term relationship that Stahl had with these communities aided my acceptance. In return, I discussed with my interlocutors how the BRP archaeological work connects to the present and is relevant to the many individuals with whom I spoke.

Interviews generally began with an introduction as to why we were there and what information we were seeking. If the informant consented (which they almost unanimously did), we began the interviews with a simple question: How have foodways changed in your lifetime or since that of your parents? The initial responses usually followed one of several patterns: (1) that no change had occurred, everything was the same, but there were different foods for different seasons; (2) that there were several dishes that were no longer made and crops no longer grown; or (3) that there were a lot of wild leaves that had formerly been used (which often were still used). Generally I continued my questions based on this first response, and let the mood and interest of the subject guide my next questions. Every interview ended with two queries: What was the biggest change (besides those already talked about) that you have seen in your lifetime or since that of your parents?; and did the interviewee have any questions for me? These open-ended questions often involved the subject asking more about the research and its value to the community, adding information they thought relevant but we had not covered, or describing changes that had had big impacts on their lives. The "biggest changes" question helped me to understand the broader social and economic changes people had seen in their lifetimes; food intersects with most of them in interesting ways.

Intensive participant observation focused primarily on three households, one in Banda, one in Bui, and one in Ngre, each representing a different socioeconomic stratum. Repeated visits were made to each household for several hours in duration to observe food preparation practices from start to finish. In addition, observation occurred during afternoon interviews, when women were busy processing various products (calabashes, kapok fibers, shea butter, etc.) and preparing food. Shorter observations occurred at random as we were passed by someone cooking or processing plants, which often led to interviews. Because interviews and observation occurred primarily during the wet season, several dry-season plant processing activities, such as sorghum processing, were not observed directly; however, several informants from different villages were questioned regarding these in an attempt to capture the variability present.

Ethnobotanical plant collection was also a focus of my 2009 ethnoarchaeological season. I focused on collecting useful wild and domesticated plants. Whenever possible I collected duplicates of the entire plant (leaves, inflorescence, seed, and roots) in order to facilitate identification and ensure appropriate material for building a phytolith comparative collection. Plant collection took place with help of community experts on wild plants; men generally knew of fruits and women of wild leaves. Both were knowledgeable about plants with other uses (for ash, soap, etc.). All plants were pressed when possible, though in practice drying presented some difficulties as it was the wet season. Consequently there was more loss than I had anticipated. I donated a duplicate set of voucher specimens to the Ghana Herbarium, Department of Botany, University of Ghana, where Dr. Patrick Ekpe kindly identified them.

I also attempted to the best of my abilities to record words for plants, tools, and dishes in Nafaanra and Kuulo/Dumpo. Transcription was phonetic, where possible, based on the spellings used by the Nafaanra Literacy Project and the knowledge of my research assistant, Enoch Mensah. These spellings are imperfect, but where possible I have cross-referenced them with the Nafaanra dictionary available online as well as with linguistic work by Blench among the Dumpo/Kuulo (Blench 2007).

ARCHIVAL METHODS

Archival work was essential for augmenting understanding of the colonial period. Building on the detailed archival work of both Stahl (2001) and Cruz (2003; Stahl and Cruz 1998), I conducted research in the Ghana National Archives in Accra, Kumasi, and Sunyani, which provided insight into colonial perceptions of Banda in the first half of the twentieth century, and also anchored many of my informants' memories in calendar years. Though archival information is sparse, it provides tantalizing glimpses of Banda life as viewed through the eyes of others.

In 2009 I visited the Brong-Ahafo Regional Archives in Sunyani which houses information from approximately the late colonial to independence eras (around 1925 to 1970s), as well as the Kumasi branch, which focuses on the period when Banda was considered part of the Ashanti region (1900–1920s). The Sunyani archives were examined in 2009 with the help of Enoch Mensah; all Banda regional material was examined, as well as documents on agricultural policy affecting Brong-Ahafo.

In 2011 I returned for a more intensive look at the Accra and Kumasi archives. The Accra archive contained a mix of material from all periods. There I examined material relating to food and agriculture in the colonial era, excepting that of cocoa and oil palm, which was the majority. I also reexamined archival sources cited in Stahl (2001) and Cruz (2003), which led me to several lists of foods, trade, and taxations, as well as colonial descriptions of Banda. Another important source of data were trade records from Wenchi and Kintampo, where Banda inhabitants traveled to ply their wares. In the Kumasi archives, all material relating to the Banda region was examined. Additional archival material from the colonial era is available at the National Archives (UK), which have been surveyed previously by Stahl (2001), and were not consulted further for this study.

APPENDIX B

Archaeobotanical Data

TABLE 3 Summary Botanical Dataset

	Late Ngre / Early Kuulo	Mid-Late Kuulo	Early Makala	Late Makala	Totals
Contexts	5	5	4	4	18
Samples	44	129	77	31	281
Total liters collected	160	605	188	142	1095
Charcoal count	6,998	21,802	7,117	10,051	45,968
Charcoal weight	107	367	192	211	877
Unidentifiable Plant Remains					
Nonseed plant remains	1,292	2,778	1,291	1,182	6,543
Seed fragments	134	357	498	142	1,131
Nut/fruit shells	8	3	1	18	30
Identified seeds	217	371	5,743	127	6,458
Total Counts	8,649	25,311	14,650	11,520	60,130

TABLE 4 Ubiquity Analysis, by Mound and Context

Phase	Site	Mound identification	No. of samples	*Pennisetum glaucum* (incl. cf.)[*]	*Sorghum bicolor* (incl. cf.)	*Zea mays* (cupule + kernel + cf.)	*Byproduct*
Late Ngre / Early Kuulo	KK	138	4	0	0		
	KK	101, L. 20–25	6	100	17		
	NK	3	3	100	67		
	NK	6	21	48	5		
	KK	148	9	75	0		
	NK	8	2	50	0		
Total/Avg. for phase			*45*	*56*	*4*	*—*	*—*
Mid-Late Kuulo	KK	101	15	93	0	0	40
	NK	8	6	100	0	0	17
	NK	7 lower	26	19	4	0	15
	KK	118	55	35	5	15	5
	NK	7 upper	25	48	0	0	32
Total/Avg. for phase			*127*	*44*	*3*	*6*	*17*
Early Makala	MK	5	49	43	33	10	
	MK	6	15	20	13	0	
	NK	8	5	60	20	20	
	A-212	1	4	100	50	0	
	B-112	1	3	100	0	100	
Total/Avg. for phase			*76*	*45*	*28*	*12*	*—*
Early/Late Makala	Banda Rockshelter	9	9	11	0	22	
Late Makala	BK	6	6	0	0	33	
	MK	9	8	13	0	38	
	MK	10	8	13	0	0	
Total/avg. for phase			*22*	*9*	*0*	*23*	*—*

Abbreviations: NK: Ngre Kataa (site); KK: Kuulo Kataa (site); MK: Makala Kataa (site); BK: Bui Kataa (site); lower: structure that is stratigraphically lower; upper: structure that is stratigraphically above the lower structure.

[*] This and the next three columns report the percentage of samples in a given context where the indicated taxon was recovered; for example, 100 indicates that 100% of samples yielded the taxon. To calculate ubiquity, the number of samples with a given taxon is divided by the total number of samples and multiplied by 100 to generate a percentage. Totals for each phase are averages for the phase, calculated by dividing the total number of samples in that phase with that taxon by the total number of samples in the phase and multiplying by 100. The pros and cons of ubiquity analysis are further considered in chapter 2, backnote 9.

TABLE 5 Summary Count and Percent Frequency Data for Grain Crops

Phase and site	Mound identification	Count: Pearl millet (Pennisetum glaucum) (incl. cf.)	%: Pearl millet	Count: Sorghum (Sorghum bicolor) (incl. cf.)	%: Sorghum	Count: cf. Digitaria	%: Digitaria	Count: Maize (Zea mays) (cupule + kernel + cf.)	%: Maize	Count: All grains
Late Ngre/Early Kuulo										
KK	138	0	0	0	0	0	0	0	0	0
KK	101, L. 20–25	44	98	1	2	0	0	0	0	45
NK	3	30	91	3	9	0	0	0	0	33
NK	6	35	95	2	5	0	0	0	0	37
KK	148	15	100	0	0	0	0	0	0	15
NK	8, L. 10–29	34	100	0	0	0	0	0	0	34
Total/avg. for phase		*158*	*96*	*6*	*4*	*0*	*0*	*0*	*0*	*164*
Mid-Late Kuulo										
KK	101	42	100	0	0	0	0	0	0	42
NK	8	8	80	1	10	0	0	1	0	10
NK	7 lower	24	92	2	8	0	0	0	0	26
KK	118	43	67	4	6	0	0	17	27	64
NK	7 upper	27	100	0	0	0	0	0	0	27
Total/avg. for phase		*144*	*85*	*7*	*4*	*0*	*0*	*18*	*11*	*169*
Early Makala										
MK	5	2,631	55	2,098	44	0	0	17	0	4,746
MK	6	3	60	2	40	0	0	0	0	5
A-212	1	12	55	10	45	0	0	0	0	22
B-112	1	4	31	0	0	0	0	9	69	13
Total/avg. for phase		*2,650*	*55*	*2,110*	*44*	*0*	*0*	*26*	*1*	*4,786*
Late Makala										0
BK	6	0	0	0	0	0	0	13	100	13
Banda Rockshelter	9	1	33	0	0	0	0	2	67	3
MK	9	1	20	0	0	0	0	4	80	5
MK	10	1	100	0	0	0	0	0	0	1
Total/avg. for phase		*3*	*14*	*0*	*0*	*0*	*0*	*19*	*86*	*22*

Note: Two kinds of data are displayed in this table: simple seed counts and the percentage frequency of a grain relative to other grains. The latter measure is calculated by dividing the count of a particular grain taxon (e.g., pearl millet) for a particular context (e.g., KK M138) by the total number of grains (pearl millet, sorghum, maize, and *Digitaria*) for

(*Contd.*)

TABLE 5 *(Continued)*

that same context (e.g., KK M138) and multiplying the result by 100 to provide a percentage. The resulting number gives us an idea of the importance of a particular grain (e.g., pearl millet) relative to all the other grains. Phase averages make the same calculation by dividing the total count of a given grain taxon by the total number of all grains for the phase.

Abbreviations: NK: Ngre Kataa (site); KK: Kuulo Kataa (site); MK: Makala Kataa (site); BK: Bui Kataa (site); lower: structure that is stratigraphically lower; upper: structure that is stratigraphically above the lower structure.

TABLE 6 Presence/Absence of Plant Taxa, by Phase

Taxon						
Dactylonium aegypticum	1		+			w
Digitaria cf. *exilis*	1					
Pennisetum glaucum	20	+	+	+		+
cf. *Pennisetum glaucum*	15	+	+	+	+	
Sorghum bicolor	11	+	+	+		
cf. *Sorghum bicolor*	6	+	+	+		
Zea mays	7		+	+	+	+
cf. *Zea mays*	1		+			
Poaceae (probable domesticates)	18	+	+	+		+
Nongrass taxa						
cf. *Abelmoschus esculentus*	2					+
Adansonia digitata	1			+		
Afromamum melegueta	2	+	+			
Butyrospermum parkii (shell)	7	+		+		
cf. *Capsicum* sp.	1					+
cf. *Cassia occidentalis*	1					+
Cassia tora	3		+	+		+
Ceiba pentandra	1					
Celtis integrifolia	2	+				+
Elaeis guineensis (shell)	1		+			
cf. *Euphorbia* sp.	1		+			
Ficus sp.	8	+	+	+		+
cf. *Indigofera tinctoria*	1					+
cf. *Laportea aestivans*	1		+			
Nicotiana cf. *rustica*	3–5	+ ?	+	+		+
Piliostigma thonningii	1		+			
Portulaca foliosa	2			+		+
Portulaca sp.	2			+		+
Sida sp.	1					+
Vigna unguiculata	1				+	
cf. *Vigna unguiculata*	4	+	+			
Zaleya pentandra	8	+	+			+
Apocynaceae	1				+	
cf. Asteraceae	1		+			
Boraginaceae	3	+	+			
Boraginaceae/Euphorbiaceae	1		+			

(Contd.)

TABLE 6 (Continued)

Grain crops and grasses	Mound contexts*	Late Ngre / Early Kuulo	Mid-Late Kuulo	Early Makala	Early/Late Makala (Banda Rockshelter)	Late Makala
No. of samples		45	132	72	9	22
Chenopodiaceae/Amaranthaceae	6		+	+		+
Cucurbitaceae	2					+
cf. Cucurbitaceae	1					
Cyperaceae/Polygonaceae	1			+		
Euphorbiaceae/Lamiaceae	5	+	+	+		
Euphorbiaceae/Malvaceae	1			+		
Fabaceae	8	+	+		+	+
cf. Fabaceae	6	+	+	+		+
Lamiaceae (Cassia/Ocimum)	1		+			
Malvaceae indeter. (cf. Sida sp.)	3	+	+		+	
Polygonaceae	2			+		+
cf. Solanaceae	2			+		+
cf. Verbenaceae	1			+		
Unknown Type 58	1				+	

* Indicates the number of broadly defined contexts in which a taxon was found. See Logan and Stahl (2017) for more detail on how mound contexts were defined.

Wild Leaves Used by Modern Villages

TABLE 7 Wild Leaves Used, by Village

Nafaanra/Twi name	Scientific name/Gloss						
				X		X	
Ayoyo						X	
Bobnondele		X		X	X		
Bobono		X					
Bokoboko		X					
Bombo	Cassia occidentalis	X				X	
Bonku/Gbonku		X		X	X		
Chasigbɔɔ	Ocimum gratissimum			X	X		X
Diagyakunu		X			X		
Diko				X			
Floe/Froe	Cissus populnea	X	X	X	X	X	X
Flewe	Pavetta crassipes			X		X	
Flongbe		X					
Flonye	Corchorus tridens	X	X	X	X	X	X
Fɔɔli	Corchorus olitorius	X	X	X		X	
Fumbɛ	Vitex doniana	X				X	

(Contd.)

TABLE 7 *(Continued)*

		Banda-Ahenkro	Bui	Dorbor	Dumpofie	Makala	Nyiire
Gabuwere	*Wild onion*	X			X		
Gbanja flonye		X		X	X	X	
Gboli				X			
Gbulo		X		X			X
Hanyini jlon	*Trema orientalis*						
Jamanyiri	*Moringa sp.*			X			
Jangboro		X				X	
Jelija	*Hibiscus cannabinus*	X	X	X		X	X
Jyakuno				X	X	X	
Jyangburu				X	X		X
Kafapoe/ kafapuwee				X		X	
Kalameshia		X			X		
Kambgee				X			
Kemeshiama		X					
KlaklokagbƐƐ	*Laportea aestuans*			X	X		
KlandƐƐ	*Cissus sp.*	X			X	X	X
Klododu	*Uvaria chamae*						
Kokoyerere				X	X	X	
Koo						X	
Kpankpan	*cf. Hymecardia acida*						
Kputukputu		X					
Kuka/Gbongbo	*Adansonia digitata*	X	X	X	X	X	
Kusumpura				X			
Lakro	*Bridelia feruginea*						
LƆm		X	X	X	X	X	
LƆmbaoe/Longbowere		X		X			
Mbandele							X
Moliere					X		
Nafaawere		X					

TABLE 7 *(Continued)*

		Banda-Ahenkro	Bui	Dorbor	Dumpofie	Makala	Nyiire
Napu	*Ocimum basilicum*	X		X	X	X	X
Ndele		X					
Ngasingasi					X		
Ngbrengbre				X			
Ngoli				X	X	X	
Ngunocho-kambele		X		X			
Nom			X				
Nyanya		X					X
NyiƐƐ		X	X	X		X	X
Mbabluwi		X					
Peelolo	*Parinari curatellifolia*						
Shia	*Ficus gnaphalocarpa*					X	
Sugbosemna						X	
Sulom				X			
Takatumo		X					
Yomaa	*Piliostigma thonningii*						
Zinblekole		X					
Zizibi				X			
Total leaves named, by village		32	8	31	20	23	11

Identifications were limited to those leaves that could be collected (i.e., were in season and still available) and dried. Identifications were made by Dr. Patrick Ekpe, of the Botany Department and Herbarium of University of Ghana, Legon.

INTRODUCTION

1. I use the term "Africa" throughout this book purposefully when referring to stereotypes about the continent, since that is the unit of analysis most people adopt when making generalizations. Africanists have long argued that the continent's cultural and geographic diversity cannot be flattened into a singular unit like "Africa" (Curtin 1969; Keim 2014). Most scholars, including myself, use more specific geographical and cultural terms when speaking of actual data and case studies, a convention I follow in this book.

2. See the work of the Past Global Changes project's LandCover6k working group, at http://pastglobalchanges.org/science/wg/landcover6k/intro, accessed June 2020.

CHAPTER 1. EXCAVATING *LONGUE DURÉE* HISTORIES
OF FOOD SECURITY IN AFRICA

1. Peggy Nelson and colleagues (2016) have developed vulnerability frameworks to understand shifts in food and other securities over time. Their work has focused on the American Southwest, where there is both a long history of detailed archaeology work and ideal preservation that affords a wealth of high-quality data and excellent visibility of past activities. This is not the case for many archaeological settings, particularly in the African continent. My approach relies on a more qualitative, mixed methods view of the past that takes advantage of the nature of archaeological data but also highlights food preference, an aspect often overlooked in more quantitative analyses. Ideally, we will soon be at a point in Africa where we can assess vulnerability in the past more clearly.

2. These implications are not straightforward. Most archaeologists associate the beginnings of food production with a more predictable food supply and greater efficiency in the food quest. But Lee (1968) suggests that foragers can provide sufficient food with less labor

than agriculturalists (see also Sahlins 1968). Many have critiqued his model, but the idea of an affluent forager does raise the interesting question of whether the advent of agriculture improved food security outcomes, particularly since agriculturalists also tended to develop social hierarchy that could diminish food access for some. His model also highlights the high nutritive value of wild resources (e.g. *mongongo* nuts) that are still used and desired by many African communities. See de Luna (2016) for a more nuanced critique of surplus and hunter-gatherers.

3. I have made a concerted effort to use *food* or *food systems* instead of *subsistence* to refer to African food and farming. Many scholars prefer *subsistence* because it acknowledges that plants and animals are used in a wider array of activities than simply provisioning. This is true, but the word also carries judgmental overtones, with the implication that "subsisting" is barely getting by. *Subsistence* is also commonly used when describing food systems, like agriculturalists, as evolutionary types rather than simply strategies used in different contexts. Given that I am trying to get away from these kinds of stereotypes I adopt food-related terms, which are also more widely used outside archaeology.

4. These time periods are rooted in chronologies devised by the Banda Research Project, which are based on changes in ceramic styles as well as AMS dates, discussed in more detail by Stahl (1999b, 2001, 2007). Chronologies like this are constantly being redefined as boundaries between archaeological phases can be arbitrary. I rely on the most recent archaeological phasing discussed in Logan and Stahl (2017), based on ceramic typologies and seventy-seven radiocarbon dates.

5. Taphonomy refers to what happens to material remains after their primary use, including how they came to be deposited in the archaeological record and the natural and cultural processes that affected them after deposition, including recovery by archaeologists. In short, taphonomy is the study of differential preservation.

6. Starch grain analysis was conducted for multiple time periods and contexts, but very few starch grains were recovered (Logan 2012).

7. Banda's primary local language is Nafaanra (Niger-Congo language family), spoken by about sixty-one thousand speakers in Ghana and more in Ivory Coast. The language has been intensively studied under the leadership of Dean Jordan and Attah Sampson, supported by the Summer Institute of Linguistics since the 1970s, resulting in the translation of the Bible in 2015 (www.ethnologue.com/language/nfr). I draw on this work and the associated dictionary when possible, but food and crop terms were understandably beyond the work of this undertaking. Orthography of words mentioned in the text is a best attempt and has been checked with native speakers.

CHAPTER 2. CHOOSING LOCAL OVER GLOBAL DURING
THE COLUMBIAN EXCHANGE

1. African grain crops that do have a significant market share (e.g., sorghum) are still treated as inferior to Southwest Asian/European grains, perhaps because their primary use outside Africa is for animal feed. Pearl millet, for example, is encountered in the United States only as bird seed. Presumably the use of both sorghum and pearl millet as animal feed relates to their low cost—and value—on a worldwide market as well as their superior nutritional qualities.

2. Bread is not typically regarded as a native food in Ghana but rather as a colonial import, so de Marees's description warrants some explanation here. "Bread" has in other cases been used by European observers as a gloss for a people's main food stuffs, but in this case de Marees does seem to refer very specifically to actual bread. He goes on to say: "Other Negroes, who live among the Portuguese, grind it [maize] on its own, without mixing it with *millie*, and bake good, excellent bread out of it, like Leiden buns, and do business selling it to the Portuguese. They know how to bake it in such a way that it keeps for three or four months" (de Marees 1987, 113, see also p. 40). In this context, it seems likely that the bread to which he refers was made specifically for Europeans. Elsewhere, he does mention that *millie* bread is baked on hot earth, after hot coals are removed (40, 112), a method which is similar to the preparation of flatbreads that were at the time widespread across the Sahel (Lewicki 1974; see below). The translators of de Marees's account suggest he is describing *kenkey* (de Marees 1987, 40), a fermented corn dough, but none of his descriptions of how it is made (through baking on the hot ground) or his comparisons to European bread fit this explanation. Either de Marees was mistaken about the preparation method used, or people may have been baking various kinds of breads, likely flatbreads, given pearl millet's lack of gluten as well as the hot-stone cooking method, possibly from a mixture of millet flour with limited imported wheat flour. Corn appears to have been consumed by African children roasted on the cob, and also to have been prepared into a beer (*pito*), probably for sale (de Marees 1987, 113).

3. Archaeological sites were named after the most proximate villages, with the addition of *kataa*, a Nafaanra word meaning "the dead" or "the ancestors" (Stahl 2001). Ngre is also spelled Nyiire according to the villagers who live there, which we only learned after excavating the archaeological site. I retain the name Ngre for the archaeological site, and Nyiire for the modern village.

4. All radiocarbon dates reported for Banda are calibrated. Calibration converts radiocarbon years to calendar years by accounting for changes in atmospheric carbon over time. In many cases, there is a significant difference between calibrated and uncalibrated date ranges, so uncalibrated dates should be used with caution. In this book, calibrated dates are used unless noted. The most recent phasing, reported in Logan and Stahl (2017, 1361–63), relies on twenty-one radiocarbon dates that when averaged using their 95 percent confidence interval fall between AD 1414 and 1615. Specific contexts may date to more specific ranges within that wider period, or may be slightly earlier or later than the period, but the vast majority will fall within that range. I have used a somewhat broader range in this section, because the specific dates we have for maize trend slightly later than the phase dates.

5. Each mound represents an area of repeated deposition, sometimes in the form of structures built atop one another or trash deposited atop earlier middens. These deposits can be accumulated rapidly or over centuries, a process archaeologists evaluate based on the style of ceramics found within as well as numerous radiocarbon dates.

6. The high degree of fragmentation makes it difficult to distinguish individual domesticates, so instead we can more securely identify them by size class. Size 2 and 3 mammals are likely sheep and goats and other similarly sized creatures, while sizes 4 and 5 include domesticated cattle and large wild animals. The data suggest a stronger reliance on sheep and goat than cattle (Logan and Stahl 2017, 1386–87).

7. In a monocropped grain field, harvesting would have commenced through beating the dried grains off the stalk, or more likely, cutting down all of the plants and threshing. Either method would have resulted in the harvesting of weedy grasses alongside domesticated grains (see D'Andrea and Casey 2002).

8. Pearl millet seeds are tremendously variable in their size and morphology, since the domesticated crop can interbreed freely with weedy forms (D'Andrea and Casey 2002). To date, there are no archaeological criteria for separating early- from late-maturing varieties, and the diversity of the species in general may prevent varietal classification.

9. The benefit of ubiquity measures is that they help account for differences in raw counts that may be due to accidents or taphonomic factors (e.g., two grains of pearl millet may not mean twice as much was used, just that the cook was sloppy that day). However, ubiquity measures are particularly sensitive to how context is defined. Most archaeobotanists use individual flotation samples to define context, since this is the most spatially and temporally delimited context available. For Banda, most flotation samples were taken as "scatter" samples. That is, a representative sample was taken from the exposed context, most commonly a 10 cm level within a 2x2m unit. Some samples are meant to target specific use contexts, like hearth contents. These contextual details are reported in appendix B and Logan (2012). The disadvantage of using such small context units is that counts are typically quite low; to account for this effect, Logan and Stahl (2017) compile multiple flotation samples into larger contexts (e.g., a single mound dating to the mid- to late Kuulo phase) for the purposes of comparison to faunal remains. Those results are different than reported here, with much higher ubiquities overall, but confirm the general trends reported here.

10. Average ubiquity by phase is calculated by determining the number of samples within a taxa divided by the number of samples for the entire phase.

11. Brackets are translators' notes in the original.

12. *N. tabacum* was grown from Mexico southwards and would have been the crop accessible to the Portuguese, who were the first to establish trade networks on Ghana's coast.

13. This is a commonly made assumption in archaeology that has been around at least since the time of V. Gordon Childe (1936). However it is problematic in that craft specialists are arranged in a wide variety of ways. Full-time farmers may produce certain crafts in the off-season, for example.

CHAPTER 3. TASTING PRIVILEGE AND PRIVATION DURING ASANTE RULE AND THE ATLANTIC SLAVE TRADE

1. Ayesha Harruna Attah, "Slow Cooking History," *New York Times*, November 11, 2018, https://www.nytimes.com/2018/11/10/opinion/sunday/slow-cooking-history.html.

2. Social evolutionary thought was predominant in archaeology at the time of Wilks's writing, and continues to be alluring to some archaeologists as well as the general public today. This kind of framework emphasizes the evolution of societies into different politico-subsistence units. Hunter-gatherers, for example, are assigned tribe- or band-level social organization. Their subsistence mode requires a certain degree of mobility that often prevents sedentism and limits population growth. Consequently, there is little impetus for the development of hereditary hierarchy. Agriculture, on the other hand, is associated with sedentary populations and increasing population densities, leading to the emergence of

hereditary leaders (Service 1962). Archaeologists have studied the transitions between these imagined types for decades, though research questions have grown more nuanced. While social evolutionary thought removes some of the progressive evolutionary baggage associated with nineteenth-century thought, some similarities remain, like the idea that agricultural state level societies are more advanced than those occupying a lower rung of the evolutionary ladder.

3. Some historians argue that Wilks's argument is mostly about labor—and the tremendous amount of it needed, he estimates, to clear primary forest—but it builds on a long tradition of scholarship (ultimately rooted in colonial stereotypes) of seeing ancient people as unable to penetrate the forest (chapter 4). His reasoning is rooted in part in the commonly accepted argument that African agriculture is limited by a general scarcity of labor and abundance of land (Hopkins 1973). It is well known that Asante and earlier coastal societies before them imported labor in exchange for gold in the sixteenth and seventeenth centuries, before turning into a supplier of enslaved labor in the eighteenth century (Lovejoy 2011). Yet the long history of agriculture and reasonably high population densities, as in Banda, Begho, and Bono Manso, suggest that if anything, the importation of labor helped certain locations to intensify production (of agriculture or gold, we do not know) rather than provided the initial push for early cultivation (see also Austin 2005).

4. In some ways, the attribution of Western economic "rationality" to African farmers pushes back against other stereotypes of Africans as irrational. On the other side, this kind of reasoning perpetuates the view that African agriculture was incapable of supporting local populations or producing a surplus, an interpretation which undermines claims to local food sovereignty today and largely recreates the biases of global agri-development (chapters 5 and 6). It is also factually incorrect based on the evidence presented in this book.

5. Pellagra or *kwashiorkor* is a disease caused by niacin and trypophan deficiency, and is most commonly found in people who consume too much maize. Through the process of nixtamalization, lime is added to maize and niacin is made nutritionally available, preventing pellagra.

6. He refers specifically to the campaign against the Fantes in 1806; Banda was also part of this campaign (Stahl 2001, 156).

7. In this parenthetical citation I believe Bowdich is referring to their, i.e., the Asante's, "corn," which from this description is most likely pearl millet. The reference to grains not cleared from the husk must refer to pearl millet; the husks of sorghum are far too hard to be edible, and maize seeds are not enclosed within individual husks and it cannot be consumed without removing the husk from outside the cob, which means this is probably not what he refers to. Pearl millet husks are much softer and are difficult to remove in processing. Pearl millet was also commonly made into a "*kouskous*" in modern and medieval (Lewicki 1974) West Africa, either by steaming or simply roughly grinding raw millet on a stone and adding a bit of water (the latter is the *sisa* referred to above). Bowdich is contrasting this "corn" of the Asantes with that of the Fantes, which seems to refer to maize given the reference to *kenkey*. The Asantes eat maize not as *kenkey*, but directly off the cob after roasting, based on his description.

8. Early Makala is too late to date effectively using AMS techniques, so dates have been established based on the cross-dating of imports, on two thermoluminescence dates, and on oral historical accounts (Stahl 1999b, 13).

9. At the time Pearsall conducted the analysis (the mid-1990s), there was not yet a single archaeological application of phytolith analysis in the African continent, hence identification methods designed for other ecological contexts were employed as a means of preliminary assessment.

10. There are two TL dates: one from Mound 5, Unit 4E2S, Level 3 yielded a date of 190± 15 (AD 1790–1820 at 1σ, 1775–1835 at 2σ); another from Mound 6, Unit 0W24N, Level 6 yielded a date of 225± 30 (AD 1740–1800 at 1σ, 1710–1830 at 2σ). Datable imports confirm this date range (Stahl 1999b, 13).

11. These taxa may also simply be compound weeds; I base the consumption of green soup upon analogy with the present, but I may be in error doing so. It is possible that regular consumption of soup made from wild herbs and greens developed later during the period of food insecurity that followed, c. 1830–1890.

12. There are charred chunks of sorghum and millet that are clearly part of articulated heads in these units.

13. There is a much higher ratio of byproduct to clean grains in these units (see figure 3).

14. These grindstones may well be associated with food preparation, though it is notable that there are absolutely no seeds recovered from this unit. My ethnographic observations in Dorbour, a potting village west of the Banda hills studied previously by Cruz (2003, 2011), suggest that grinding stones as well as several other bodily logics used in food preparation are used for preparation of pastes. Regarding the gender of the potter, it is of course always possible that potters were previously male, but all ethnographic and ethnohistorical data suggest that potters in this region were women (Stahl and Cruz 1998).

15. Based on the observation that plant remains are much better preserved and are present in large quantities in this area compared to elsewhere, this seems to be a locus of the conflagration (also Stahl 2001, 169).

16. Banda Rockshelter is also known by its site name, B-2 (see map 2). It is difficult to date when the cave was actually used. The pottery is sufficiently generic to the Makala phase, and prevents defining any specific Early or Late Makala associations (Stahl, personal communication 2012). Stahl (2001, 157–58) reiterates the association in oral histories with the cave and hiding from the Asante, but we do not know if this refers to initial conflicts (1773/74) or later ones during the nineteenth century. We may also not be dealing with Banda peoples at all; Stahl (2001, 150) cites Agbodeka (1971), who mentioned that Nkoranza people hid in the Banda hills during a dispute with Asante in 1893, though it is unclear if they used this specific cave. However the pottery is similar to that at Makala Kataa (Stahl 2001, 157)—while we wouldn't want to assign a modern ethnic label to this it seems most parsimonious that it represents people who were associated with Banda area sites.

17. Leith Smith excavated two 1x2m units as part of a regional testing program (BRP 2002).

CHAPTER 4. CREATING CHRONIC FOOD INSECURITY IN THE GOLD COAST COLONY

1. Establishing a baseline referent "fixes" a certain point in time, even if elements of multiple times are often conflated into a single reference point (Stahl 1993). The use of a baseline also collapses time and change. The millennia captured by the label "precolonial" spanned

a considerable number of political, economic, cultural, and environmental shifts that were no less complex than the periods for which we have extensive written documentation, as is documented by a wide array of historians, archaeologists, and other specialists on these periods. The nature of our sources on precolonial history means that our chronologies are less refined than in the familiar documentary sources, which can create the appearance of more stasis even if that is not the case (Stahl 2001, 20–27).

2. No linguistic data are available for the precolonial period in this region. While Shaffer admits data are scarce for the precolonial period, he argues for precolonial hunger based on linguistic data from sources dating to later periods. His earliest sources come from two early colonial-era reports. The first is an 1899 report from Captain Henry Northcott which "is incomplete and contains errors but is nevertheless indicative" (Shaffer 2017, 274). Northcott notes that a term for "hunger" appears in seven local dialectics in the North (*kom* or a derivative in Dagomba, Mamprusi, Moshi, Wa, and Gurunsi; and *erkun* or *ekon* in Daboya and Bole). The second source, District Commissioner A.W. Cardinall's 1906 report, includes a transcription of terms in Kasem and Sissala compiled by missionaries in Navrongo, which did discriminate hunger, *kana*, from starvation, *kananga*. Rattray's much later (1932) work is also cited as providing words for hunger and starvation in thirteen languages. Note that none of these sources specifically points to a hungry season or seasonal hunger. Instead, Shaffer apparently concludes that there was a pronounced hunger season based on the distinctions made in modern Mole-Dagbani languages (citing a report by Wazam in his personal possession), as well as from mid- to late twentieth-century proverbs and greetings. At best, these data hint at a wide understanding of hunger and its distinction from starvation by the late nineteenth century, which makes sense given the context of widespread upheaval. This is interesting and important, but does not address the issue of the hunger season, or of hunger in precolonial times. It should be emphasized that none of this data is derived from historical comparative methods that may be able to trace the concept into precolonial times.

3. These laterite floors exposed by York appear to be associated with a nineteenth-century artifact assemblage (imported pipes, bullets, gunflints, glazed sherds), though it is quite possible that other contexts he excavated, including a destruction layer that included the skeleton mentioned below, predate this occupation. People may also have reinhabited homes built in previous decades.

4. ARG 1/20/7. This is but one example of several archival sources that mention low population density and social and economic "backwardness" in Banda in the early twentieth century; see Stahl (2001, 192–94, 197–98) for a more comprehensive account.

5. In the Late Makala phase, maize is the most commonly encountered grain, but ubiquity is still quite low (19%) compared to previous phases, when grain ubiquity percentages were usually in the 40s for the most prevalent grain. It is hard to determine whether the lack of charred grains is a real pattern or a postdepositional one. Late Makala deposits present a particular challenge for archaeological plant remains, for there were no midden mounds and domestic structures were shallowly buried (Stahl 2001, 200). Indeed, the domestic structure sampled, at Station 10, revealed only one grain of pearl millet. But other charred seeds of non grain seeds, usually less well preserved in the Banda sequence, were recovered, which supports the idea that grains were simply not as prominently used. A large pit at Makala Kataa Station 9 provides a better record of culinary and economic activities.

This rectangular pit was about a meter deep and had been filled in with refuse (Stahl 2001, 203). While pearl millet counts remained low, there was more maize present and in more samples. Okra was also present, as were wild leafy greens/ruderals. Given the consistency in patterning (more maize than millet, no sorghum) at sites with different depositional contexts, such as Bui Kataa and Banda Rockshelter, I suspect the patterns described here are real rather than purely postdepositional.

6. Grinding stones may also be rare at Late Makala occupations of Makala Kataa if they were removed and relocated to the new village site constructed as part of a colonial rebuilding scheme in the late 1920s (Stahl 2001, 196–97).

7. I have investigated the possibility that maize was one such cash crop that might have been encouraged by British colonial policies, but have no evidence that this was the case in the early colonial period. Some West African colonies did export maize during the first decade of the twentieth century, notably Togo and Nigeria, but Ghana did not, and it appears to have become a net importer of maize by the mid-twentieth century (Miracle 1966). However there is the possibility that maize may have been encouraged to supply internal markets, although we have no data on this dynamic for the early colonial period.

8. One exception was during the invasion of locusts, which colonial officials did their best to monitor because locusts had such devastating impacts on local harvests that locust invasions brought some communities to the brink of famine. Shortly before 1930, colonial records were abuzz with attempts to control locusts near Banda, all the while complaining that the natives were too lazy to do what was needed to staunch their pursuit. Yet to hear the elderly recall locust invasions conjures quite a different perspective. They say that tuber crops are little affected, since their growth is underground, and that locust attacks are random—they do not attack everyone's fields uniformly. When this topic came up in interviews, every old person I talked to instantly smiled, and recounted how "sweet" or delicious locust meat was. Apparently, people's response to locusts was not to prevent their arrival, as colonial officials had recommended, but to welcome it with open arms so that they could capture thousands of teeming locusts and feast upon them. This is a classic example of what Richards (1985) calls working with nature rather than against it.

9. Purcell's first report, focusing on Akim, is published in a 1939 volume I was able to access. However subsequent drafts of the report are available only in the Accra archives. I did not encounter them in the archives in 2011, so I rely on Robins as a detailed secondary source for their content. Note that these reports should have been published, as was the first one, but they were suppressed and not meant to be accessible.

10. Plange (1979, 7–8) suggests that this ritual or special view of cattle was a relatively recent development at the time; since northern cattle traders had been terminated by the colonial system, making cattle more of a special than an everyday thing.

11. Sanitation was the justification for massive colonial projects in the Gold Coast, which included among other things the destruction of old African villages and the relocation of their inhabitants to new, more "sanitary" villages. These so-called sanitary villages were built along grids of wide streets, facilitating easy surveillance. Banda oral histories are replete with references to the *bruni* (white foreign) "breaker of walls." Banda villages were destroyed and relocated in the late 1920s, which is also why our archaeological record extends only to this time; all later occupations are buried under modern habitations (Stahl 2001). The colonial obsession with sanitation and hygiene was one of the most pervasive impacts of colonial rule on people's daily lives, effectively disciplining African bodies into

Western ways of living and doing. These echoes of this process persist in the Banda youth's insistence that "olden times" foods must be repackaged in modern, sanitary packages for people to wish to purchase them (chapter 6).

CHAPTER 5. CONSUMING A REMOTELY GLOBAL MODERNITY IN RECENT TIMES

1. Ages reported in this chapter refer to the time of interviewing, 2009.

2. *Fnumu* are seeds of *Lagenaria siceraria*, the bottle gourd. The wild form is native to southern Africa but was domesticated very early in the Americas. It most likely diffused back to the African continent in its domesticated form through the Columbian Exchange. Peanuts are also a South American domesticate. Along with maize, these crops are excellent indicators of how Banda's cuisine has been global in some sense for centuries. No firm identifications of *fnumu* or peanut were made in archaeological contexts, but this is likely a preservation bias rather than an indication of absence in the past.

3. Sieving may be a recently added step in *tuo zafi* preparation to accommodate the less uniform texture of flour from the diesel mills, or it may have always been needed to accommodate a diversity of inclusions in flour.

4. See appendix C for the full listing of leaves from interviews across five villages. Sixty-three different names were recorded, and while I attempted to minimize repetition it is possible that different names may refer to the same plant.

5. I acknowledge Gabrielle Hecht for coining this phrase and encouraging the analysis it provoked in a graduate class, "Bodies, Natures, and Technologies in Africa," at University of Michigan.

CHAPTER 6. EATING AND REMEMBERING PAST CULTURAL ACHIEVEMENTS

1. Medical anthropologists in particular have long recognized this dynamic in medical interventions upon African people; archaeologists would do well to explore their cautionary tales and critical engagements.

2. The Banda Cultural Centre was built beginning in the mid-1990s and again in the early 2000s by community members, supported by funds obtained by Ann Stahl from the National Science Foundation. Since then, its large structures have served as a repository for artifacts and a hub for archaeological lab work in the years when excavations were ongoing, as a place to stay for archaeologists as well as visitors to the area, and as a community meeting space for large gatherings.

3. While the event itself consumed only a day, about six weeks were spent planning, designing, and consulting about the event. Prior to the event, Enoch and I presented topical posters to several interest groups, including the paramount chief and his (male) elders, and the queen mother and her (female) elders, in order to prompt conversation and also discuss what should be covered at the event itself. The original focus of the event, archaeological and historical research that had been conducted by Ann Stahl and myself, was greatly expanded to include a variety of crafts that local craftspeople still made (pottery, baskets, etc.) or made until recently (cloth) as well as performances of local music traditions and examples of traditional ceremonial dress. We also did one-on-one trainings with "historical

ambassadors," individuals hand-selected for their knowledge of and interest in the past, drawing from my interviewees as well as the highly skilled excavators from the Banda Research Project. These individuals acted as translators and tour guides for each poster topic during the event. Following the event, focus groups were invited to continue conversation on the most promising topics, one of which was heritage foods.

4. Focus groups were recruited during and immediately after the Heritage Day event. At the main event, local guides explained posters associated with several themes, including food. Particularly interested and engaged individuals were invited to participate in focus groups after the main event. These focus groups ranged from six to twenty individuals.

5. See Hamza Moshood, "Colonialism Walks into a Chop Bar," *Popula*, September 24, 2019, https://popula.com/2019/09/27/colonialism-walks-into-a-chop-bar/.

REFERENCES

ARCHIVAL SOURCES

Brong Ahafo Regional Archives

BRG 28/2/5. Banda Native Affairs, 1922–1931.

Ghana National Archives, Accra

ADM 56/1/153. Report on Tour of Inspection in the Northern Territories by the Director of Agriculture, 1912.

ADM 56/1/415. Monthly Reports, Kintampo District for 1901.

ADM 56/1/421. Monthly Reports, Kintampo District for 1906.

ADM 56/1/423. Annual Report, Kintampo District for 1906.

ADM 56/1/458. Annual Report, Kintampo District for 1902.

Kumasi Archives

ARG 1/20/7. Report of Recent Tour of Inspection in the Northern District of Ashanti, 1907.

ARG 1/2/21/3. Banda Native Affairs, 1922–1940. (Fell's 1913 report on Banda history is reproduced within this document by H.J. Hobbs to honor a request by A.F.L. Wilkinson.)

ORAL HISTORIES

Stahl, Ann B., and J. M. Anane. 2011. *Family Histories from the Banda Traditional Area, Brong-Ahafo Region, Ghana.* Manuscript on file with the Institute of African Studies and Department of Archaeology, University of Ghana, and Ghana National Museums.

ARCHAEOLOGICAL FIELD NOTES

Banda Research Project (BRP)

1989   Makala Kataa. Excavation Notes.

1990   Makala Kataa, Brong-Ahafo. Excavation Notes, Stations 6, 9, 10.

1994   Makala Kataa. Excavation Notes, Stations 6 and 10.

1995   Kuulo Kataa. Excavation Notes.

2000   Kuulo Kataa. Excavation Field Notes.

2002   Regional Testing Field Notes.

2008   Bui Kataa and Ngre Kataa. Field Notebooks.

2009   Ngre Kataa. Field Notes.

OTHER SOURCES

Abbiw, D.K. 1990. *Useful Plants of Ghana*. London: Intermediate Technology Publications and Royal Botanic Gardens, Kew.

Acemoglu, Daron, S. Johnson, and James A. Robinson. 2002. "Reversal of Fortune: Geography and Institutions in the Making of the Modern World." *Quarterly Journal of Economics* 117 (4): 1231–94.

Akyeampong, Emmanuel Kwaku. 1996. *Drink, Power, and Cultural Change: A Social History of Alcohol in Ghana, c. 1800 to Recent Times*. Oxford: Heinemann.

Akyeampong, Emmanuel Kwaku, Robert Bates, Nathan Nunn, and James A Robinson, eds. 2014. *Africa's Development in Historical Perspective*. Cambridge: Cambridge University Press.

Albala, Ken, ed. 2014. *The Food History Reader: Primary Sources*. London: Bloomsbury.

Allman, Jean, and Victoria Tashjian. 2000. *"I Will Not Eat Stone": A Women's History of Colonial Asante*. Portsmouth, NH: Heinemann.

Alpern, Stanley B. 1992. "The European Introduction of Crops into West Africa in Precolonial Times." *History in Africa* 19 (1992): 13–43.

———. 2008. "Exotic Plants of Western Africa: Where They Came From and When." *History in Africa* 35: 63–102.

Ameyaw, K. 1965. "Traditions of Banda." In *Traditions from Brong-Ahafo, Nos. 1–4*, 1–15. Legon: Institute of African Studies, University of Ghana.

Andah, Bassey. 1995. "Studying African Societies in Cultural Context." In *Making Alternative Histories: The Practice of Archaeology and History in Non-Western Settings*, edited by Peter R. Schmidt and Thomas Patterson, 149–82. Santa Fe: School of American Research Press.

Arhin, Kwame. 1967a. "The Financing of the Ashanti Expansion (1700–1820)." *Africa* 37 (3): 283–91.

———. 1967b. "The Structure of Greater Ashanti (1700–1824)." *The Journal of African History* 8 (1): 65–85.

———. 1970. "Aspects of the Ashanti Northern Trade in the Nineteenth Century." *Africa: Journal of the International African Institute* 40 (4): 363–73.

———. 1974. *The Papers of George Ekem Ferguson: A Fanti Official of the Government of the Gold Coast, 1890–1897*. Cambridge: African Studies Centre.

———. 1983. "Peasants in 19th-Century Asante." *Current Anthropology* 24 (4): 471–80.

———. 1987. "Savanna Contributions to the Asante Political Economy." In *Golden Stool: Studies of the Asante Center and Periphery*, edited by Enid Schildkrout, 51–59.

Anthropological Papers of the American Museum of Natural History, 65, part 1. New York: American Museum of Natural History.

———. 1990. "Accumulation and the State in Asante in the Nineteenth Century." *Africa* 60 (4): 524–37.

Arhin, Kwame, and J. Ki-Zerbo. 1989. "States and Peoples of the Niger Bend and the Volta." In *General History of Africa*. Vol. 6, *Africa in the Nineteenth Century until the 1880s*, edited by J. Ajayi, 662–98. Berkeley: UNESCO.

Austin, Gareth. 2005. *Labour, Land and Capital in Ghana: From Slavery to Free Labour in Asante, 1807–1956.* Rochester, NY: University of Rochester Press.

———. 2007. "Labour and Land in Ghana: 1874–1939: A Shifting Ratio and an Institutional Revolution." *Australian Economic History Review* 47 (1): 95–120.

———. 2008a. "Resources, Techniques, and Strategies South of the Sahara: Revising the Factor Endowments Perspective on African Economic Development, 1500–2000." *Economic History Review* 61 (3): 587–624.

———. 2008b. "The 'Reversal of Fortune' Thesis and the Compression of History: Perspectives from African and Comparative Economic History." *Journal of International Development,*: 996–1027.

Austin, Gareth, Jorg Baten, and Alexander Moradi. 2007. "Exploring the Evolution of Living Standards in Ghana, 1880–2000: An Anthropometric Approach." Working Papers 7021, Economic History Society.

Austin, Gareth, Jorg Baten, and Bas van Leeuwen. 2012. "The Biological Standard of Living in Early Nineteenth-Century West Africa: New Anthropometric Evidence for Northern Ghana and Burkina Faso." *Economic History Review* 65 (4): 1280–1302.

Ball, Terry, Karol Chandler-Ezell, Ruth Dickau, Neil Duncan, Thomas C. Hart, Jose Iriarte, Carol Lentfer, et al. 2016. "phytoliths as a Tool for Investigations of Agricultural Origins and Dispersals around the World." *Journal of Archaeological Science* 68: 32–45.

Baro, Mamadou, and Tara Deubel. 2006. "Persistent Hunger : Perspectives on Vulnerability, Famine, and Food Security in Sub-Saharan Africa." *Annual Review of Anthropology* 35: 521–38.

Bassil, Noah R. 2011. "The Roots of Afropessimism: The British Invention of the 'Dark Continent.'" *Critical Arts* 25 (3): 377–96.

Beichner, Paul E. 1961. "The Grain of Paradise." *Speculum: A Journal of Medieval Studies* 36 (2): 302–7.

Berry, Sara S. 1984. "The Food Crisis and Agrarian Change in Africa: A Review Essay." *African Studies Review* 27 (2): 59–112.

———. 1993. *No Condition Is Permanent : The Social Dynamics of Agrarian Change in Sub-Saharan Africa.* Madison: University of Wisconsin Press.

Blench, Roger. 2007. "The Dompo Language of Central Ghana and Its Affinities." Unpublished draft available at http://www.rogerblench.info/Language/Niger-Congo/Kwa/Dompo%20 Wordlist.pdf.

Boaten, Kwasi. 1970. "Trade among the Asante of Ghana to the End of 18th Century." *Institute of African Studies Research Review* 7 (1): 33–52.

Boivin, Nicole, Dorian Q. Fuller, and Alison Crowther. 2012. "Old World Globalization and the Columbian Exchange: Comparison and Contrast." *World Archaeology* 44 (3): 452–69.

Bonnecase, Vincent. 2018. "When Numbers Represented Poverty: The Changing Meaning of the Food Ration in French Colonial Africa." *Journal of African History* 59 (3): 463–81.

Boserup, Ester. 1965. *The Conditions of Agricultural Growth. The Economics of Agrarian Change under Population Pressure.* London: Allen and Unwin.

Bosman, Willem. 1705. *A New and Accurate Description of the Coast of Guinea, Divided into the Gold, the Slave, and the Ivory Coasts: Containing a Geographical, Political and Natural History of the Kingdoms and Countries.* London: J. Knapton.

Bostoen, Koen. 2007. "Pearl Millet in Early Bantu Speech Communities in Central Africa: A Reconsideration of the Lexical Evidence." *Afrika Und Übersee* 89: 183–212.

Bostoen, Koen, Bernard Clist, C. Doumenge, Rebecca Grollemund, Jean Marie Hombert, Joseph Koni Muluwa, and Jean Maley. 2015. "Middle to Late Holocene Paleoclimatic Change and the Early Bantu Expansion in the Rain Forests of Western Central Africa." *Current Anthropology* 56 (3): 354–84.

Bowdich, Thomas Edward. 1873 [1819]. *Mission from Cape Coast Castle to Ashantee; with a Statistical Account of That Kingdom, and Geographical Notices of Other Parts of the Interior of Africa.* London: Griffith and Farran.

Brunken, J., J.M.J. de Wet, and J.R. Harlan. 1977. "The Morphology and Domestication of Pearl Millet." *Economic Botany* 31 (2): 163–74.

Cardinall, A.W. 1931. *The Gold Coast, 1931: A Review of Conditions in the Gold Coast in 1931 as Compared with Those of 1921, Based on Figures and Facts Collected by the Chief Census Officers of 1931, Together with a Historical, Ethnographical and Sociological Survey of the People.* Accra: Government Printer.

Carney, Judith. 2001. *Black Rice: The African Origins of Rice Cultivation in the Americas.* Cambridge, MA: Harvard University Press.

Carney, Judith, and Richard Nicholas Rosomoff. 2009. *In the Shadow of Slavery: Africa's Botanical Legacy in the Atlantic World.* Berkeley: University of California Press.

Carr, Edward R. 2008. "Between Structure and Agency: Livelihoods and Adaptation in Ghana's Central Region." *Global Environmental Change* 18 (4): 689–99.

Casey, Joanna. 2000. *The Kintampo Complex: The Late Holocene on the Gambaga Escarpment, Northern Ghana.* Oxford: Archaeopress.

Chakrabarty, Dipesh. 2000. *Provincializing Europe: Postcolonial Thought and Historical Difference.* Princeton, NJ: Princeton University Press.

Cherniwchan, Jevan, and Juan Moreno-Cruz. 2019. "Maize and Precolonial Africa." *Journal of Development Economics* 136 (March): 137–50.

Childe, V. Gordon. 1936. *Man Makes Himself.* London: Watts.

Chouin, Gérard. 2012. "The 'Big Bang' Theory Reconsidered: Framing Early Ghanaian History." *Transactions of the Historical Society of Ghana,* 14: 13–40.

Christaller, J.G., and Kofi Ron Lange. 2000. *Three Thousand Six Hundred Ghanaian Proverbs (from the Asante and Fante Language).* Lewiston, NY: E. Mellen Press.

Clark, Gracia. 1994. *Onions Are My Husband: Survival and Accumulation by West African Market Women.* Chicago: University of Chicago Press.

Cliggett, Lisa. 2005. *Grains from Grass: Aging, Gender, and Famine in Rural Africa.* Ithaca, NY: Cornell University Press.

Cobb, Charles R. 2005. "Archaeology and the 'Savage Slot': Displacement and Emplacement in the Premodern World." *American Anthropologist* 107 (4): 563–74.

Coil, James, M. Alejandra Korstanje, Steven Archer, and Christine A. Hastorf. 2003. "Laboratory Goals and Considerations for Multiple Microfossil Extraction in Archaeology." *Journal of Archaeological Science* 30: 991–1008.

Comaroff, Jean, and John L. Comaroff. 1991. *Of Revelation and Revolution*, Vol. 1, *Christianity, Colonialism, and Consciousness in South Africa*. Chicago: University of Chicago Press.

Comaroff, John L., and Jean Comaroff. 1992. *Ethnography and the Historical Imagination*. Boulder, CO: Westview Press.

———. 1997. *Of Revelation and Revolution*, Vol. 2, *The Dialectics of Modernity on a South African Frontier*. Chicago: University of Chicago Press.

Cordell, D. 2003. "The Myth of Inevitability and Invincibility: Resistance to Slavers and the Slave Trade in Central Africa, 1850–1910." In *Fighting the Slave Trade: West Africa Strategies*, edited by S.A. Diouf, 31–49. Athens: Ohio University Press.

Counihan, Carole. 2004. *Around the Tuscan Table: Food, Family, and Gender in Twentieth-Century Florence*. New York: Routledge.

Cowan, Ruth Schwartz. 1983. *More Work for Mother: The Ironies of Household Technologies from the Open Hearth to the Microwave*. New York: Basic Books.

Crosby, Alfred W. 2003 [1972]. *The Columbian Exchange: Biological and Cultural Consequences of 1492*. Westport, CT: Praeger.

Cruz, M. Dores. 2003. "Shaping Quotidian Worlds: Ceramic Production and Consumption in Banda, Ghana c. 1780–1994." Ph.D. diss., Department of Anthropology, Binghamton University, State University of New York.

———. 2011. "'Pots Are Pots, Not People': Material Culture and Ethnic Identity in the Banda Area (Ghana), Nineteenth and Twentieth Centuries." *Azania: Archaeological Research in Africa* 46 (3): 336–57.

Curtin, Philip. 1964. *The Image of Africa: British Ideas and Action 1780–1850*. Madison: University of Wisconsin Press.

———. 1969. *The Atlantic Slave Trade: A Census*. Madison: University of Wisconsin Press.

D'Andrea, A. Catherine, and Joanna Casey. 2002. "Pearl Millet and Kintampo Subsistence." *African Archaeological Review* 19 (3): 147–73.

D'Andrea, A. Catherine, S. Kahlheber, A.L. Logan, and D.J. Watson. 2007. "Early Domesticated Cowpea (*Vigna unguiculata*) from Central Ghana." *Antiquity* 81 (313): 686–98.

D'Andrea, A. Catherine, M. Klee, and J. Casey. 2001. "Archaeobotanical Evidence for Pearl Millet (*Pennisetum glaucum*) in Sub-Saharan West Africa." *Antiquity* 75 (288): 341–48.

D'Andrea, A. Catherine, Amanda L. Logan, and Derek J. Watson. 2006. "Oil Palm and Prehistoric Subsistence in Tropical West Africa." *Journal of African Archaeology* 4 (2): 195–222.

Daoud, Adel. 2010. "Robbins and Malthus on Scarcity, Abundance, and Sufficiency: The Missing Sociocultural Element." *American Journal of Economics and Sociology* 69 (4): 1206–29.

David, Nicholas, and Carol Kramer. 2001. *Ethnoarchaeology in Action*. New York: Cambridge University Press.

Davis, Mike. 2001. *Late Victorian Holocausts: El Niño Famines and the Making of the Third World*. New York: Verso.

DeCorse, Christopher R. 2005. "Coastal Ghana in the First and Second Millennia AD: Change in Settlement Patterns, Subsistence, and Technology." *Journal des Africanistes* 75 (2): 43–54.

de Garine, Igor. 1997. "Food Preferences and Taste in an African Perspective: A Word of Caution." In *Food Preferences and Taste: Continuity and Change*, edited by Helen Macbeth, 187–99. Providence, RI: Berghahn Books.

de Luna, Kathryn M. 2016. *Collecting Food, Cultivating People: Subsistence and Society in Central Africa*. New Haven, CT: Yale University Press.

de Luna, Kathryn M., and Jeffrey B. Fleisher. 2019. *Speaking with Substance: Methods of Language and Materials in African History*. New York: Springer.

de Luna, Kathryn M., Jeffrey B. Fleisher, and Susan Keech McIntosh. 2012. "Thinking across the African Past: Interdisciplinarity and Early History." *African Archaeological Review* 29 (2–3): 75–94.

de Marees, Pieter. 1987 [1602]. *Description and Historical Account of the Gold Kingdom of Guinea*. Oxford: Published for British Academy by Oxford University Press.

DeVault, M.L. 1991. *Feeding the Family: The Social Organization of Caring as Gendered Work*. Chicago: University of Chicago Press.

Devereux, Stephen. 2001. "Famine in Africa." In *Food Security in Sub-Saharan Africa*, edited by Stephen Devereux and Simon Maxwell, 117–48. London: ITDG.

Devereux, Stephen, and Simon Maxwell, eds. 2001. *Food Security in Sub-Saharan Africa*. London: ITDG.

de Waal, Alex. 1989. *Famine That Kills: Darfur, Sudan, 1984–1985*. Oxford: Clarendon Press.

de Wet, J.M.J., and J.R. Harlan. 1971. "The Origin and Domestication of Sorghum Bicolor." *Economic Botany* 25 (2): 128–35.

Diamond, Jared M. 1999. *Guns, Germs, and Steel: The Fates of Human Societies*. New York: Norton.

Dickson, K. 1964. "The Agricultural Landscape of Southern Ghana and Ashanti Brong-Ahafo: 1800–1850." *Bulletin of the Ghana Geographical Association* 9 (1): 25–35.

Dietler, Michael. 2007. "Culinary Encounters: Food, Identity, and Colonialism." In *The Archaeology of Food and Identity*, edited by Katheryn C. Twiss, 218–42. Carbondale, IL: Center for Archaeological Investigations, Southern Illinois University.

Di Giovine, Michael A., and Ronda L. Brulotte. 2014. "Introduction: Food and Foodways as Cultural Heritage." In *Edible Identities: Food as Cultural Heritage*, edited by Ronda Brulotte and Michael Di Giovine, 1–28. Farnham, UK: Ashgate Publishing.

Douglas, Mary. 2003 [1966]. *Purity and Danger: An Analysis of Concepts of Pollution and Taboo*. New York: Routledge.

Doyle, Shane. 2006. *Crisis and Decline in Bunyoro: Population and Environment in Western Uganda 1860–1955*. Oxford: James Currey Publishers.

Dueppen, Stephen A., and Cameron Gokee. 2014. "Hunting on the Margins of Medieval West African States: A Preliminary Study of the Zooarchaeological Record at Diouboye, Senegal." *Azania* 49 (3): 354–85.

Dupuis, Joseph. 1966 [1824]. *Journal of a Residence in Ashantee, by Joseph Dupuis, . . . Comprising Notes and Researches Relative to the Gold Coast and the Interior of Western Africa . . .* London: H. Colburn.

Ehret, Christopher. 2002. *The Civilizations of Africa: A History to 1800*. Oxford: Currey.

———. 2014. "Africa in World History before ca. 1440." In *Africa's Development in Historical Perspective*, edited by E. Akyeampong, Robert Bates, Nathan Nunn, and James A. Robinson, 33–55. Cambridge: Cambridge University Press.

Escobar, Arturo. 2012. *Encountering Development: The Making and Unmaking of the Third World*. Princeton, NJ: Princeton University Press.

Esterhuysen, Amanda B., and Shannon K. Hardwick. 2017. "Plant Remains Recovered from the 1854 Siege of the Kekana Ndebele, Historic Cave, Makapan Valley, South Africa." *Journal of Ethnobiology* 37 (1): 97–119.

Fabian, Johannes. 2002. *Time and the Other: How Anthropology Makes Its Object*. New York: Columbia University Press.

Fairhead, James, and Melissa Leach. 1996. *Misreading the African Landscape: Society and Ecology in a Forest-Savanna Mosaic*. Cambridge: Cambridge University Press.

Ferguson, James. 1990. *The Anti-Politics Machine: "Development" Depoliticization, and Bureaucratic Power in Lesotho*. Cambridge: Cambridge University Press.

——. 2006. *Global Shadows: Africa in the Neoliberal World Order*. Durham, NC: Duke University Press.

Ferme, Mariane. 2001. *The Underneath of Things: Violence, History, and the Everyday in Sierra Leone*. Berkeley: University of California Press.

Field, Margaret. 1931. "Gold Coast Food." *Petit Propos Culinaires* 42: 7–21.

Fields-Black, Edda L. 2008. *Deep Roots: Rice Farmers in West Africa and the African Diaspora*. Bloomington: Indiana University Press.

Forster, Stig, Wolfgang J. Mommsen, and Ronald Robinson. 1988. *Bismarck, Europe, and Africa: The Berlin Africa Conference 1884–1885 and the Onset of Partition*. Oxford: Oxford University Press.

Forsyth, J. 1962. "Major Food Storage Problems." In *Agriculture and Land Use in Ghana*, edited by J.B. Wills, 394–401. London: Oxford University Press.

Fortes, M., and S.L. Fortes. 1936. "Food in the Domestic Economy of the Tallensi." *Africa* 9 (2): 237–76.

Freidberg, Susanne. 2003. "French Beans for the Masses: A Modern Historical Geography of Food in Burkina Faso." *Journal of Historical Geography* 29 (3): 445–63.

Gallagher, Daphne. 2016. "American Plants in Sub-Saharan Africa: A Review of the Archaeological Evidence." *Azania* 51 (1): 24–61.

Gallagher, Daphne E., Stephen A. Dueppen, and Rory Walsh. 2016. "The Archaeology of Shea Butter (*Vitellaria Paradoxa*) in Burkina Faso, West Africa." *Journal of Ethnobiology* 36 (1): 150–71.

Garcin, Yannick, Pierre Deschamps, Guillemette Ménot, Geoffroy de Saulieu, Enno Schefuß, David Sebag, Lydie M. Dupont, et al. 2018. "Early Anthropogenic Impact on Western Central African Rainforests 2,600 Y Ago." *Proceedings of the National Academy of Sciences* 115 (13): 3261–66.

Garrard, T.F. 1980. *Akan Weights and the Gold Trade*. London: Longman.

Gasser, R.E., and E.C. Adams. 1981. "Aspects of Deterioration of Plant Remains in Archaeological Sites: The Walpi Archaeological Project." *Journal of Ethnobiology* 1 (1): 182–92.

Gautier, Achilles, and W. van Neer. 2005. "The Continuous Exploitation of Wild Animal Resources in the Archaeozoological Record of Ghana." *Journal of African Archaeology* 3 (2): 195–212.

Giblin, James L. 1992. *The Politics of Environmental Control in Northeastern Tanzania, 1840–1940*. Philadelphia: University of Pennsylvania Press.

Gijanto, Liza, and Sarah Walshaw. 2014. "Ceramic Production and Dietary Changes at Juffure, Gambia." *African Archaeological Review* 31 (2): 265–97.

Gocking, Roger S. 2005. *The History of Ghana*. Westport, CT: Greenwood Press.

González-Ruibal, Alfredo. 2008. "Time to Destroy." *Current Anthropology* 49 (2): 247–79.

——. 2014. *An Archaeology of Resistance: Materiality and Time in an African Borderland*. Lanham, MD: Rowman and Littlefield.

Goody, Jack. 1977. *The Domestication of the Savage Mind*. Cambridge: Cambridge University Press.

———. 1982. *Cooking, Cuisine and Class: A Study in Comparative Sociology*. Cambridge University Press.

Goucher, Candice Lee. 1981. "Iron Is Iron 'til It Is Rust: Trade and Ecology in the Decline of West African Iron-Smelting." *Journal of African History* 22: 179–89.

Green, Toby. 2019. *A Fistfull of Shells: West Africa from the Rise of the Slave Trade to the Age of Revolution*. Chicago: University of Chicago Press.

Grier, B. 1981. "Underdevelopment, Modes of Production, and the State in Colonial Ghana." *African Studies Review* 24 (1): 21–47.

Guyer, Jane I. 1978. "The Food Economy and French Colonial Rule in Central Cameroun." *The Journal of African History* 19 (4): 577–97.

———. 1980. "Food, Cocoa, and the Division of Labor by Sex in 2 West-African Societies." *Comparative Studies in Society and History* 22 (3): 355–73.

———. 1984. "Naturalism in Models of African Production." *Man* 19: 371–88.

———. 1988. "The Multiplication of Labor: Historical Methods in the Study of Gender and Agricultural Change in Modern Africa." *Current Anthropology* 29 (2): 247–72.

———. 1995. "Wealth in People, Wealth in Things—Introduction." *Journal of African History* 36 (1): 83–90.

———. 2004. *Marginal Gains: Monetary Transactions in Atlantic Africa*. Chicago: University of Chicago Press.

Guyer, Jane I., and Samuel M. Eno Belinga. 1995. "Wealth in People as Wealth in Knowledge: Accumulation and Composition in Equatorial Africa." *Journal of African History* 36 (1): 91–120.

Ham, Jessica R. 2017. "Cooking to Be Modern but Eating to Be Healthy: The Role of Dawa-Dawa in Contemporary Ghanaian Foodways." *Food, Culture and Society* 20 (2): 237–56.

Hammond, Dorothy, and Alta Jablow. 1970. *The Africa That Never Was: Four Centuries of British Writing about Africa*. New York: Twayne Publishers.

Handloff, R. 1987. "Trade and Politics on the Asante Periphery: Gyaman, 1818–1900." In *The Golden Stool: Studies of the Asante Center and Periphery*, edited by Enid Schildkrout, 252–59. New York: American Museum of Natural History.

Hastorf, Christine A. 2003. "Andean Luxury Foods: Special Food for the Ancestors, Deities and the Elite." *Antiquity* 77 (297): 545–54.

———. 2017. *The Social Archaeology of Food: Thinking about Eating from Prehistory to Present*. New York: Cambridge University Press.

Hawthorne, Walter. 2003. *Planting Rice and Harvesting Slaves: Transformations along the Guinea-Bissau Coast, 1400–1900*. London: Heinemann.

Hegmon, Michelle, Margaret C. Nelson, and Karen Gust Schollmeyer. 2016. "Experiencing Social Change: Life during the Mimbres Classic Transformation." *Archeological Papers of the American Anthropological Association* 27 (1): 54–73.

Hendrie, Barbara. 1997. "Knowledge and Power: A Critique of an International Relief Operation." *Disasters* 21 (1): 57–76.

Hill, Polly. 1963. *The Migrant Cocoa-Farmers of Southern Ghana: A Study of Rural Capitalism*. Cambridge: Cambridge University Press.

Hobsbawm, Eric, and Terence Ranger. 1983. *The Invention of Tradition*. Cambridge: Cambridge University Press.

Hodge, Joseph Morgan. 2011. *Triumph of the Expert: Agrarian Doctrines of Development and the Legacies of British Colonialism*. Athens: Ohio University Press.

Holtzman, Jon. 2006. "Food and Memory." *Annual Review of Anthropology* 35: 361–78.

———. 2009. *Uncertain Tastes: Memory, Ambivalence, and the Politics of Eating in Samburu, Northern Kenya*. Berkeley: University of California Press.

Hopkins, A.G. 1973. *An Economic History of West Africa*. New York: Columbia University Press.

———. 2009. "The New Economic History of Africa." *The Journal of African History* 50 (2): 155–77.

House, L.R. 1995. "Sorghum and Millets: History, Taxonomy, and Distribution." In *Sorghum and Millets: Chemistry and Technology*, edited by David Dendy, 1–10. St. Paul, MN: American Association of Cereal Chemists.

Huffman, Thomas N. 1996. "Archaeological Evidence for Climatic Change during the Last 2000 Years in Southern Africa." *Quaternary International* 33: 55–60.

Hutton, William. 1821. *A Voyage to Africa: Including a Narrative of an Embassy to One of the Interior Kingdoms, in the Year 1820; with Remarks on the Course and Termination of the Niger, and Other Principal Rivers in That Country*. London: Printed for Longman, Hurst, Rees, Orme, and Brown.

Iliffe, John. 1987. *The African Poor: A History*. Cambridge: Cambridge University Press.

———. 1995. *Africans: The History of a Continent*. Cambridge: Cambridge University Press.

Inikori, J.E. 1982. *Forced Migration: The Impact of the Export Slave Trade on African Societies*. London: Hutchinson University Library for Africa.

Janer, Zilkia. 2007. "(In)Edible Nature: New World Food and Coloniality." *Cultural Studies* 21 (2–3): 385–405.

Jones, Martin, Harriet Hunt, Emma Lightfoot, Diane Lister, Xinyi Liu, and Giedre Motuzaite-Matuzeviciute. 2011. "Food Globalization in Prehistory." *World Archaeology* 43 (4): 665–75.

Jones, William O. 1959. *Manioc in Africa*. Palo Alto, CA: Stanford University Press.

Kahlheber, Stefanie, Koen Bostoen, and Katharina Neumann. 2009. "Early Plant Cultivation in the Central African Rain Forest: First Millennium BC Pearl Millet from South Cameroon." *Journal of African Archaeology* 7 (2): 253–72.

Kahlheber, Stefanie, Manfred K.H. Eggert, Dirk Seidensticker, and Hans-Peter Wotzka. 2014. "Pearl Millet and Other Plant Remains from the Early Iron Age Site of Boso-Njafo (Inner Congo Basin, Democratic Republic of the Congo)." *African Archaeological Review* 31 (3): 479–512.

Kea, Ray A. 1982. *Settlements, Trade, and Polities in the Seventeenth-Century Gold Coast*. Baltimore: Johns Hopkins University Press.

Keen, David. 1994. *The Benefits of Famine: The Political Economy of Famine and Relief in South Sudan, 1983–1989*. Princeton, NJ: Princeton University Press.

Keim, Curtis A. 2014. *Mistaking Africa: Curiosities and Inventions of the American Mind*. 3rd ed. Boulder, CO: Westview Press.

Kelly, Kenneth G. 1995. "Transformation and Continuity in Savi, a West African Trade Town: An Archaeological Investigation of Culture Change on the Coast of Benin during the 17th and 18th Centuries." Ph.D. diss., Department of Anthropology, University of California Los Angeles.

Klein, A. N. 1994a. "Reply to Wilk's Commentary on 'Slavery and Akan Origins.'" *Ethnohistory* 41: 666–67.

———. 1994b. "Slavery and Akan Origins?" *Ethnohistory* 41 (4): 627–56.

———. 1996. "Toward a New Understanding of Akan Origins." *Africa* 66 (2): 248–73.

Koenig, Dolores. 2006. "Food for the Malian Middle Class: An Invisible Cuisine." In *Fast Food/Slow Food: The Cultural Economy of the Global Food System*, edited by Richard Wilk, 49–68. Lanham, MD: AltaMira Press.

La Fleur, James Daniel. 2012. *Fusion Foodways of Africa's Gold Coast in the Atlantic Era.* Leiden: Brill.

Lane, Paul J. 2005. "Barbarous Tribes and Unrewarding Gyrations? The Changing Role of Ethnographic Imagination in African Archaeology." In *African Archaeology: A Critical Introduction*, edited by Ann B. Stahl, 24–54. Malden, MA: Blackwell.

———. 2015. "Archaeology in the Age of the Anthropocene: A Critical Assessment of Its Scope and Societal Contributions." *Journal of Field Archaeology* 40 (5): 485–98.

Lappé, Frances Moore, and Joseph Collins. 1978. "Why Can't People Feed Themselves?" In *Food First: Beyond the Myth of Scarcity*, edited by Frances M Lappe and Joseph Collins, 75–85. New York: Ballantine.

LaViolette, Adria, and Stephanie Wynne-Jones. 2018. *The Swahili World.* New York: Routledge.

Law, Robin. 1995. *From Slave Trade to "Legitimate" Commerce: The Commercial Transition in Nineteenth-Century West Africa.* Cambridge: Cambridge University Press.

Leach, Melissa, and Robin Mearns, eds. 1996. *The Lie of the Land: Challenging Received Wisdom on the African Environment.* Oxford: James Currey.

Lee, Richard. 1968. "What Hunters Do for a Living, or, How to Make Out on Scarce Resources." In *Man the Hunter*, edited by Richard Lee and Irven DeVore, 30–48. New York: Atherton Press.

Lewicki, Tadeusz. 1974. *West African Food in the Middle Ages: According to Arabic Sources.* London: Cambridge University Press.

Liu, Xinyi, Emma Lightfoot, Tamsin C. O'Connell, Hui Wang, Shuicheng Li, Liping Zhou, Yaowu Hu, Giedre Motuzaite-Matuzeviciute, and Martin K. Jones. 2014. "From Necessity to Choice: Dietary Revolutions in West China in the Second Millennium BC." *World Archaeology* 46 (5): 661–80.

Logan, Amanda L. 2006. "The Application of Phytolith and Starch Grain Analysis to Understanding Formative Period Subsistence, Ritual, and Trade on the Taraco Peninsula, Highland Bolivia." Master's thesis, Department of Anthropology, University of Missouri, Columbia.

———. 2012. "A History of Food without History: Food, Trade, and Environment in West-Central Ghana in the Second Millennium AD." Ph.D. diss., Department of Anthropology, University of Michigan, Ann Arbor.

———. 2016a. "An Archaeology of Food Security in Banda, Ghana." *Archeological Papers of the American Anthropological Association* 27 (1): 106–19.

———. 2016b. "Why Can't People Feed Themselves? Archaeology as Alternative Archive of Food Security in Banda, Ghana." *American Anthropologist* 118 (3): 508–24.

———. 2017. "Will Agricultural Technofixes Feed the World? Short- and Long-Term Tradeoffs of Adopting High-Yielding Crops." In *The Give and Take of Sustainability: Archaeological and Anthropological Perspectives*, edited by Michelle Hegmon, 109–24. Cambridge: Cambridge University Press.

Logan, Amanda L., and M. Dores Cruz. 2014. "Gendered Taskscapes: Food, Farming, and Craft Production in Banda, Ghana in the Eighteenth to Twenty-First Centuries." *The African Archaeological Review* 31 (2): 203–31.

Logan, Amanda L., and A. Catherine D'Andrea. 2012. "Oil Palm, Arboriculture, and Changing Subsistence Practices during Kintampo Times (3600–3200 BP, Ghana)." *Quaternary International* 249: 63–71.

Logan, Amanda L., and Ann B. Stahl. 2017. "Genealogies of Practice in and of the Environment in Banda, Ghana." *Journal of Archaeological Method and Theory* 24 (4): 1356–99.

Logan, Amanda L., Daryl Stump, Steven T. Goldstein, Emuobosa Akpo Orijemie, and M.H. Schoeman. 2019. "Usable Pasts Forum: Critically Engaging Food Security." *African Archaeological Review* 36 (3): 419–38.

Lorimer, Douglas A. 1978. *Class, Colour, and the Victorians: A Study of English Attitudes towards the Negro in the Mid-Nineteenth Century*. New York: Holmes and Meier.

Lovejoy, Paul E. 2011. *Transformations in Slavery: A History of Slavery in Africa*. Cambridge: Cambridge University Press.

Lyons, Diane. 2007. "Integrating African Cuisines: Rural Cuisine and Identity in Tigray, Highland Ethiopia." *Journal of Social Archaeology* 7 (3): 346–71.

———. 2014. "Perceptions of Consumption: Constituting Potters, Farmers, and Blacksmiths in the Culinary Continuum in Eastern Tigray, Northern Highland Ethiopia." *African Archaeological Review* 31 (2): 169–201.

MacBeth, H., and S. Lawry. 1997. "Food Preferences and Taste: An Introduction." In *Food Preferences and Taste: Continuity and Change*, edited by Helen Macbeth, 1–14. Providence, RI: Berghahn Books.

MacEachern, Scott. 2018. *Searching for Boko Haram: A History of Violence in Central Africa*. New York: Oxford University Press.

Madella, Marco, Carla Lancelotti, and Juan José García-Granero. 2016. "Millet Microremains: An Alternative Approach to Understand Cultivation and Use of Critical Crops in Prehistory." *Archaeological and Anthropological Sciences* 8 (1): 17–28.

Maggs, Tim. 1982. "Mgoduyanuka: Terminal Iron Age Settlement in the Natal Grasslands." *Annals of the Natal Museum* 25: 83–115.

Malkki, Liisa H. 2015. *The Need to Help: The Domestic Arts of International Humanitarianism*. Raleigh-Durham, NC: Duke University Press.

Malthus, Thomas. 1798. *An Essay on the Principle of Population*. London: J. Johnson.

Mandala, Elias Coutinho. 2005. *The End of Chidyerano: A History of Food and Everyday Life in Malawi, 1860–2004*. Portsmouth, NH: Heinemann.

Manning, Patrick. 2014. "African Population, 1650–2000: Comparisons and Implications of New Estimates." In *Africa's Development in Historical Perspective*, edited by E. Akyeampong, R. Bates, N. Nunn, and J. Robinson, 131–52. Cambridge: Cambridge University Press.

Marshall, Fiona, and Elisabeth Hildebrand. 2002. "Cattle before Crops: The Beginning of Food Production in Africa." *Journal of World Prehistory* 16 (2): 99–143.

Maxwell, Simon. 2001. "The Evolution of Thinking about Food Security." In *Food Security in Sub-Saharan Africa*, edited by Stephen Devereux and Simon Maxwell, 13–31. London: ITDG.

Mayshar, Joram, Omar Moav, Zvika Neeman, and Luigi Pascali. 2015. "Cereals, Appropriability, and Hierarchy." Centre for Economic Policy Research Discussion Paper Series: Development Economics, Economic History, Macroeconomics and Growth and Public Economics, 10742. London.

McCann, James C. 1999. *Green Land, Brown Land, Black Land: An Environmental History of Africa, 1800–1990*. Portsmouth, NH; London: Heinemann; Eurospan.

———. 2001. "Maize and Grace: History, Corn, and Africa's New Landscapes, 1500—1999." *Comparative Studies in Society and History* 43 (2): 246–72.

———. 2005. *Maize and Grace: Africa's Encounter with a New World Crop, 1500–2000*. Cambridge, MA: Harvard University Press.

———. 2009. *Stirring the Pot: A History of African Cuisine*. Athens: Ohio University Press.

McCaskie, T.C. 1995. *Asante: State and Society in African History*. Cambridge: Cambridge University Press.

McIntosh, Roderick. 2005. *Ancient Middle Niger: Urbanism and the Self-Organizing Landscape*. Cambridge: Cambridge University Press.

McIntosh, Roderick J., and Susan Keech McIntosh. 1981. "The Inland Niger Delta before the Empire of Mali: Evidence from Jenne-Jeno." *The Journal of African History* 22 (1): 1–22.

———. 1984. "The Early City in West Africa: Towards an Understanding." *The African Archaeological Review* 2 (1): 73–98.

McIntosh, Susan Keech. 1995. "Pottery." In *Excavations at Jenne-Jeno, Hmbarketolo, and Kaniana (Inland Niger Delta, Mali), the 1981 Season*, edited by S.K. McIntosh, 130–213. Berkeley: University of California Press.

———, ed. 1999. *African Perspectives on Political Complexity*. New York: Cambridge University Press.

McIntosh, Susan Keech, Daphne Gallagher, and Roderick McIntosh. 2003. "Tobacco Pipes from Excavations at the Museum Site, Jenne, Mali." *Journal of African Archaeology* 1 (2): 171–99.

McLeod, Malcolm D. 1987. "Gifts and Attitudes." In *The Golden Stool: Studies of the Asante Center and Periphery*, edited by Enid Schildkrout, 184–91. New York: American Museum of Natural History.

Meskell, Lynn. 2011. *The Nature of Heritage: The New South Africa*. New York: Wiley-Blackwell.

Miller, Brandi Simpson. 2015–16. "Foodways and Empire in 19th-Century Asante Diplomatic Relations." *The SOAS Journal of Postgraduate Research* 9: 34–46.

Minc, Leah D. 1986. "Scarcity and Survival: The Role of Oral Tradition in Mediating Subsistence Crises." *Journal of Anthropological Archaeology* 5 (1): 39–113.

Mintz, Sidney W., and Christine M. Du Bois. 2002. "The Anthropology of Food and Eating." *Annual Review of Anthropology* 31 (1): 99–119.

Miracle, Marvin P. 1965. "The Introduction and Spread of Maize in Africa." *The Journal of African History* 6 (1): 39–55.

———. 1966. *Maize in Tropical Africa*. Madison: University of Wisconsin Press.

Mitchell, Peter. 2005. *African Connections: An Archaeological Perspective on Africa and the Wider World*. Lanham, MD: AltaMira Press.

Mitchell, Timothy. 2002. *Rule of Experts: Egypt, Techno-Politics, Modernity*. Berkeley: University of California Press.

Monroe, J. Cameron, and Anneke Janzen. 2014. "The Dahomean Feast: Royal Women, Private Politics, and Culinary Practices in Atlantic West Africa." *The African Archaeological Review* 31 (2): 299–337.

Moore, Henrietta L., and Megan Vaughan. 1994. *Cutting Down Trees: Gender, Nutrition, and Agricultural Change in the Northern Province of Zambia, 1890–1990*. Portsmouth, NH: Heinemann.

Moyo, Sam, Praveen Jha, and Paris Yeros. 2013. "The Classical Agrarian Question: Myth, Reality and Relevance Today." *Agrarian South: Journal of Political Economy* 2 (1): 93–119.

Mudimbé, Valentin Yves. 1994. *The Idea of Africa*. Bloomington; London: Indiana University Press; James Currey.

Murray, Shawn Sabrina. 2005. "The Rise of African Rice Farming and the Economic Use of Plants in the Upper Middle Niger Delta (Mali)." Ph.D. diss., Department of Anthropology, University of Wisconsin, Madison.

National Research Council. 1996. *Lost Crops of Africa*, Vol. 1, *Grains*. Washington, DC: National Academy Press.

Nelson, Margaret C., Scott E. Ingram, Andrew J. Dugmore, Richard Streeter, Matthew A. Peeples, Thomas H. McGovern, Michelle Hegmon, Jette Arneborg, Keith W. Kintigh, Seth Brewington, Katherine A. Spielmann, Ian A. Simpson, Colleen Strawhacker, Laura E.L. Comeau, Andrea Torvinen, Christian K. Madsen, George Hambrecht, Konrad Smiarowski. 2016. "Climate Challenges, Vulnerabilities, and Food Security." *Proceedings of the National Academy of Sciences* 113 (2): 298–303.

Neumann, Katharina. 2005. "The Romance of Farming: Plant Cultivation and Domestication in Africa." In *African Archaeology: A Critical Introduction*, edited by Ann Brower Stahl, 249–75. Malden, MA: Blackwell Publishing.

Neumann, Roderick P. 1995. "Ways of Seeing Africa: Colonial Recasting of African Society and Landscape in Serengeti National Park." *Ecumene* 2 (2): 149–169.

Niemeijer, D. 1996. "The Dynamics of African Agricultural History: Is It Time for a New Development Paradigm?" *Development and Change* 27 (1): 87–110.

Norman, Neil L. 2009. "Hueda (Whydah) Country and Town: Archaeological Perspectives on the Rise and Collapse of an African Atlantic Kingdom." *International Journal of African Historical Studies*.

Nunn, Nathan. 2008. "The Long-Term Effects of Africa's Slave Trades." *Quarterly Journal of Economics* 123 (1): 139–76.

Ogundiran, Akinwumi. 2002. "Of Small Things Remembered: Beads, Cowries, and Cultural Translations of the Atlantic Experience in Yorubaland." *The International Journal of African Historical Studies* 35 (2): 427–57.

Ohadike, D.C. 1981. "The Influenza Pandemic of 1918–19 and the Spread of Cassava Cultivation on the Lower Niger: A Study in Historical Linkages." *The Journal of African History* 22 (3): 379–91.

Ong, Aihwa. 1987. *Spirits of Resistance and Capitalist Discipline: Factory Women in Malaysia*. Albany: State University of New York Press.

Osseo-Asare, Fran. 2005. *Food Culture in Sub-Saharan Africa*. Westport, CT: Greenwood Press.

Owusuh, E.S.K. 1976. *Oral Traditions of Sampa, Hani, Debibi, Namasa, Banda, Broahene, and Mengye (Menji)—Brong-Ahafo*. Legon: Institute of African Studies, University of Ghana.

Ozanne, Paul. 1969. "The Diffusion of Smoking in West Africa." *Odu Journal of West African Studies* 1 (2): 29–42.

Pavanello, Mariano. 2015. "Foragers or Cultivators? A Discussion of Wilks's 'Big Bang' Theory of Akan History." *Journal of West African History* 1 (2): 1–26.

Pearsall, Deborah M. 2015. *Paleoethnobotany: A Handbook of Procedures*. New York: Routledge.

Pearsall, Deborah M., Karol Chandler-Ezell, and James A. Zeidler. 2004. "Maize in Ancient Ecuador: Results of Residue Analysis of Stone Tools from the Real Alto Site." *Journal of Archaeological Science* 31: 423–42.

Person, Yves. 1968. *Samori: Une Revolution Dyula*. Dakar: IFAN.

Phillips, J.E. 1983. "African Smoking and Pipes." *Journal of African History* 24 (3): 303–19.

Pierre, Jemima. 2012. *The Predicament of Blackness: Postcolonial Ghana and the Politics of Race*. Chicago: University of Chicago Press.

Pikirayi, Innocent. 2002. *The Zimbabwe Culture: Origins and Decline of Southern Zambezian States*. Walnut Creek, CA: Rowman Altamira.

———. 2006. "The Demise of Great Zimbabwe, AD 1420—1550: An Environmental Reappraisal." In *Cities in The World 1500-2000*, edited by A. Green and R. Leech, 31–47. Leeds: Maney.

Piot, Charles. 1999. *Remotely Global: Village Modernity in West Africa*. Chicago: University of Chicago Press.

———. 2010. *Nostalgia for the Future: West Africa after the Cold War*. Chicago: University of Chicago Press.

Piperno, Dolores R. 2006. *Phytoliths: A Comprehensive Guide for Archaeologists and Paleoecologists*. Lanham, MD: AltaMira Press.

Plange, N.-K. 1979. "Underdevelopment in Northern Ghana: Natural Causes or Colonial Capitalism?" *Review of African Political Economy* 15/16: 4–14.

Pomeranz, Kenneth. 2000. *The Great Divergence: China, Europe, and the Making of the Modern World Economy*. Princeton, NJ: Princeton University Press.

Popkin, Barry M., Linda S. Adair, and Shu Wen Ng. 2013. "Now and Then: The Global Nutrition Transition: The Pandemic of Obesity in Developing Countries." *Nutrition Review* 70 (1): 3–21.

Posnansky, Merrick. 1987. "Prelude to Akan Civilization." In *The Golden Stool: Studies of the Asante Center and Periphery*, edited by E. Schildkrout, 14–22. Anthropological Papers of the American Museum of Natural History, 65, part 1. New York: American Museum of Natural History.

Purcell, F.M. 1939. *Diet and Ill-Health in the Forest Country of the Gold Coast*. London: H.K. Lewis & Co.

Radomski, Kai Uwe, and Katharina Neumann. 2011. "Grasses and Grinding Stones: Inflorescence Phytoliths from Modern West African Poaceae and Archaeological Stone Artefacts." In *Windows on the African Past: Current Approaches to African Archaeobotany*, edited by Ahmed Fahmy, Stefanie Kahlheber, and A. Catherine D'Andrea, 153–66. Frankfurt: Africa Magna Verlag.

Ranger, Terence. 1976. "Towards a Usable African Past." In *African Studies since 1945: A Tribute to Basil Davidson*, edited by Christopher Fyfe, 17–30. Edinburgh: Longman.

Rattray, Robert S. 1923. *Ashanti*. Oxford: Oxford University Press.

———. 1932. *The Tribes of the Ashanti Hinterland*. Vol. 1. Oxford: Clarendon Press.

Ribot, Jesse. 2014. "Cause and Response: Vulnerability and Climate in the Anthropocene." *Journal of Peasant Studies* 41 (5): 667–705.

Richard, François. 2017. "'Excessive Economies' and the Logics of Abundance: Genealogies of Wealth, Labor, and Social Power in Pre-Colonial Senegal." In *Abundance: the Archaeology of Plenitude*, edited by Monica L. Smith, 201–28. Boulder: University Press of Colorado.

Richards, Audrey. 1964 [1932]. *Hunger and Work in a Savage Tribe: A Functional Study of Nutritional among the Southern Bantu.* Cleveland: Meridian Books.

———. 1995 [1939]. *Land, Labour, and Diet in Northern Rhodesia.* Münster: Lit.

Richards, P. 1985. *Indigenous Agricultural Revolution: Ecology and Food Production in West Africa.* Boston: Unwin Hyman.

Robertshaw, Peter. 1994. "Archaeological Survey, Ceramic Analysis, and State Formation in Western Uganda." *African Archaeological Review* 12: 105–31.

———. 2000. "Sibling Rivalry? The Intersection of Archaeology and History." *History in Africa* 27: 261–86.

Robertson, G.A. 1819. *Notes on Africa: Particularly Those Parts Which Are Situated between Cape Verd and the River Congo . . .* London: Sherwood, Neely and Jones.

Robin, Cynthia. 2013. *Everyday Life Matters: Maya Farmers at Chan.* Gainesville: University of Florida Press.

Robins, Jonathan E. 2018. "'Food Comes First': The Development of Colonial Nutritional Policy in Ghana, 1900–1950." *Global Food History* 4 (2): 168–88.

Rock, Joeva. 2018. "Abject Lessons." *Popula*, Aug. 8. https://popula.com/2018/08/08/abject-lessons/.

———. 2019. "'We Are Not Starving': Challenging Genetically Modified Seeds and Development in Ghana." *Culture, Agriculture, Food and Environment* 41 (1): 15–23.

Roddick, Andrew P., and Ann B. Stahl. 2016. *Knowledge in Motion: Constellations of Learning across Time and Place.* Tucson: University of Arizona Press.

Rodney, Walter. 1972. *How Europe Underdeveloped Africa.* London: Bogle-L'Ouverture Publications.

Rönnbäck, Klas, and Dimitrios Theodoridis. 2019. "African Agricultural Productivity and the Transatlantic Slave Trade: Evidence from Senegambia in the Nineteenth Century." *The Economic History Review* 72 (1): 209–322.

Sahlins, Marshall. 1968. "Notes on the Original Affluent Society." In *Man the Hunter*, edited by Richard Lee and Irven DeVore, 85–89. New York: Atherton Press.

Said, Edward W. 2003. *Orientalism.* London: Penguin Books.

Santos, Boaventura de Sousa. 2018. *The End of the Cognitive Empire: The Coming of Age of Epistemologies of the South.* Durham, NC: Duke University Press.

Schmidt, Peter R., and Innocent Pikirayi. 2016. *Community Archaeology and Heritage in Africa: Decolonizing Practice.* New York: Routledge.

Schoenbrun, David L. 1998. *A Green Place, a Good Place: Agrarian Change and Social Identity in the Great Lakes Region to the 15th Century.* Oxford: James Currey Publishers.

Scholliers, Peter, and Kyri W. Claflin. 2012. "Introduction: Surveying Global Food Historiography." In *Writing Food History: A Global Perspective*, edited by Kyri W. Claflin and Peter Scholliers, 1–10. London: Berg.

Scott, James C. 1977. *The Moral Economy of the Peasant: Rebellion and Subsistence in Southeast Asia.* New Haven, CT: Yale University Press.

Sen, Amartya. 1981. *Poverty and Famines: An Essay on Entitlement and Deprivation.* Oxford: Oxford University Press.

Service, Elman. 1962. *Primitive Social Organization: An Evolutionary Perspective.* New York: Random House.

Shaffer, Paul. 2017. "Seasonal Hunger in the Northern Territories of the Gold Coast, 1900–40." *Economic History of Developing Regions* 32 (3): 270–300.

Shanahan, T.M., J.T. Overpeck, K.J. Anchukaitis, J.W. Beck, J.E. Cole, D.L. Dettman, J.A. Peck, C.A. Scholz, and J.W. King. 2009. "Atlantic Forcing of Persistent Drought in West Africa." *Science* 324 (5925): 377–80.

Shinnie, Peter. 1996. "Early Asante: Is Wilks Right?" In *The Cloth of Many Colored Silks: Papers on History and Society Ghanaian and Islamic in Honor of Ivor Wilks*, edited by John O. Hunwick and Nancy Lawler, 195–203. Evanston, IL: Northwestern University Press.

———. 2005. "Early Asante and European Contacts." *Journal des Africanistes* 75 (2): 25–42.

Smith, J.N.L. 2008. "Archaeological Survey of Settlement Patterns in the Banda Region, West-Central Ghana: Exploring External Influences and Internal Responses in the West African Frontier." Ph.D. diss., Department of Anthropology, Syracuse University.

Staff Division of Agriculture. 1962. "Crops Other than Cocoa and the Diseases and Pests Which Affect Them." In *Agriculture and Land Use in Ghana*, edited by J. Brian Wills, 353–93. London: Oxford University Press.

Stahl, Ann Brower. 1985. "Reinvestigation of Kintampo 6 Rock Shelter, Ghana: Implications for the Nature of Culture Change." *African Archaeological Review* 3: 117–50.

———. 1986. "Early Food Production in West Africa: Rethinking the Role of the Kintampo Culture." *Current Anthropology* 27 (5): 532–36.

———. 1991. "Ethnic Styles and Boundaries: A Diachronic Case Study from West-Central Ghana." *Ethnohistory* 38 (3): 250–75.

———. 1993. "Concepts of Time and Approaches to Analogical Reasoning in Historical Perspective." *American Antiquity* 58 (2): 235–60.

———. 1999a. "Perceiving Variability in Time and Space: The Evolutionary Mapping of African Societies." In *Beyond Chiefdoms: Pathways to Complexity in Africa*, edited by Susan K. McIntosh, 39–55. Cambridge: Cambridge University Press.

———. 1999b. "The Archaeology of Global Encounters Viewed from Banda, Ghana." *African Archaeological Review* 16 (1): 5–81.

———. 2001. *Making History in Banda: Anthropological Visions of Africa's Past*. Cambridge: Cambridge University Press.

———. 2002. "Colonial Entanglements and the Practices of Taste: An Alternative to Logocentric Approaches." *American Anthropologist* 104 (3): 827–45.

———. 2005. "Introduction: Changing Perspectives on Africa's Pasts." In *African Archaeology: A Critical Introduction*, edited by Ann B. Stahl, 1–23. Malden, MA: Blackwell.

———. 2007. "Entangled Lives: The Archaeology of Daily Life in the Gold Coast Hinterlands, AD 1400–1900." In *Archaeology of Atlantic Africa and the African Diaspora*, edited by Akinwumi Ogundiran and Toyin Falola, 49–76. Bloomington: Indiana University Press.

———. 2008a. "Dogs, Pythons, Pots, and Beads: The Dynamics of Shrines and Sacrifical Practices in Banda, Ghana, 1400–1900 CE." In *Memory Work: Archaeologies of Material Practices*, edited by Barbara J. Mills and William H. Walker, 159–86. Santa Fe: School for Advanced Research.

———. 2008b. "The Slave Trade as Practice and Memory: What Are the Issues for Archaeologists?" In *Invisible Citizens: Captives and Their Consequences*, edited by Catherine M. Cameron, 25–56. Salt Lake City: University of Utah Press.

———. 2013. "Archaeological Insights into Aesthetic Communities of Practice in the Western Volta Basin." *African Arts* 46 (3): 54–67.

————. 2015. "Metalworking and Ritualization: Negotiating Change through Improvisational Practice in Banda, Ghana." *Archeological Papers of the American Anthropological Association* 26 (1): 53–71.

————. 2016. "Complementary Crafts: The Dynamics of Multicraft Production in Banda, Ghana." In *Gendered Labor in Specialized Economies: Archaeological Perspectives on Male and Female Work*, edited by Sophia E. Kelly and Traci Ardren, 157–88. Boulder: University of Colorado Press.

————. 2018a. "Market Thinking: Perspectives from Saharan and Atlantic West Africa." In *Market as Place and Space of Economic Exchange: Perspectives from Archaeology and Anthropology*, edited by Hans Peter Hahn and Geraldine Schmitz, 152–79. Oxford: Oxbow Books.

————. 2018b. "Efficacious Objects and Techniques of the Subject: 'Ornaments' and Their Depositional Contexts in Banda, Ghana." In *Relational Identities and Other-Than-Human Agency in Archaeology*, edited by Eleanor Harrison-Buck and Julia Hendon, 197–236. Boulder: University of Colorado Press.

Stahl, Ann Brower, and M. Dores Cruz. 1998. "Men and Women in a Market Economy: Gender and Craft Production in West Central Ghana ca. 1775–1995." In *Gender in African Prehistory*, edited by Susan Kent, 205–26. Landham, MD: Rowman and Littlefield.

Stahl, Ann Brower, Maria Das Dores Cruz, Hector Neff, Michael D. Glascock, Robert J. Speakman, Bretton Giles, and Leith Smith. 2008. "Ceramic Production, Consumption and Exchange in the Banda Area, Ghana: Insights from Compositional Analyses." *Journal of Anthropological Archaeology* 27 (3): 363–81.

Stahl, A.B., and P.W. Stahl. 2004. "Ivory Production and Consumption in Ghana in the Early Second Millennium AD." *Antiquity* 78: 86–101.

Stephens, Rhiannon. 2016. "'Wealth,' 'Poverty,' and the Question of Conceptual History in Oral Contexts: Uganda from c. 1000 C.E." In *Doing Conceptual History in Africa*, edited by Axel Fleisch and Rhiannon Stephens, 21–48. New York: Berghahn Books.

————. 2018a. "Bereft, Selfish, and Hungry: Greater Luhyia Concepts of the Poor in Precolonial East Africa." *American Historical Review* 123 (3): 789–816.

————. 2018b. "Poverty's Pasts: A Case for *Longue Durée* Studies." *Journal of African History* 59 (3): 399–409.

Stone, Glenn Davis, Robert McC. Netting, and M. Priscilla Stone. 1990. "Seasonality, Labor Scheduling, and Agricultural Intensification in the Nigerian Savanna." *American Anthropologist* 92 (1): 7–23.

Stump, Daryl. 2010. "'Ancient and Backward or Long-Lived and Sustainable?' The Role of the Past in Debates Concerning Rural Livelihoods and Resource Conservation in Eastern Africa." *World Development* 38 (9): 1251–62.

Sutton, David E. 2001. *Remembrance of Repasts: An Anthropology of Food and Memory*. Oxford and New York: Berg.

Sutton, Inez. 1989. "Colonial Agricultural Policy: The Non-Development of the Northern Territories of the Gold Coast." *The International Journal of African Historical Studies* 22 (4): 637–69.

Swanepoel, Natalie. 2005. "Socio-Political Change on a Slave-Raiding Frontier: War, Trade and 'Big Men' in Nineteenth Century Sisalaland, Northern Ghana." *Journal of Conflict Archaeology* 1 (1): 265–93.

Taylor, Delia A., P.T. Robertshaw, and R.A. Marchant. 2000. "Environmental Change and Political Economic Upheveal in Pre-Colonial Western Uganda." *The Holocene* 10 (4): 527–36.

Thomas, Nicholas. 1991. *Entangled Objects: Exchange, Material Culture, and Colonialism in the Pacific.* Cambridge, MA: Harvard University Press.

Thomas, Roger G. 1972. "George Ekem Ferguson: Civil Servant Extraordinary." *Transactions of the Historical Society of Ghana* 13 (2): 181–215.

Thompson, E.P. 1971. "The Moral Economy of the English Crowd during the Eighteenth Century." *Past and Present* 50: 76–116.

Thurow, Roger. 2012. *The Last Hunger Season: A Year in an African Farm Community on the Brink of Change.* New York: Public Affairs.

Tilley, Helen. 2011. *Africa as a Living Laboratory: Empire, Development, and the Problem of Scientific Knowledge, 1870–1950.* Chicago: University of Chicago Press.

Trouillot, Michel-Rolph. 1991. "Anthropology and the Savage Slot: The Poetics and Politics of Otherness." In *Recapturing Anthropology*, edited by Richard G. Fox, 17–44. Santa Fe: School of American Research.

———. 1995. *Silencing the Past: Power and the Production of History.* Boston: Beacon.

Trubek, Amy. 2009. *The Taste of Place: A Cultural Journey into Terroir.* Berkeley: University of California Press.

Van Esterik, Penny. 2006. "From Hunger Foods to Heritage Foods: Challenges to Food Localizations in Lao PDR." In *Fast Food/Slow Food: The Cultural Economy of the Global Food System*, edited by Richard Wilk, 83–96. Lanham, MD: AltaMira.

Van Harten, A.M. 1970. "Melegueta Pepper." *Economic Botany* 24 (2): 208–16.

Vansina, Jan. 1990. *Paths in the Rainforests: Toward a History of Political Tradition in Equatorial Africa.* Madison: University of Wisconsin Press.

———. 1995. "Historians, Are Archaeologists Your Siblings?" *History in Africa* 22: 369–408.

Vivian, Brian. 1992. "Sacred to Secular: Transitions in Akan Funerary Customs." In *An African Commitment*, edited by Judy Sterner and Nicholas David, 157–67. Calgary: Calgary University Press.

Walshaw, Sarah C. 2010. "Converting to Rice: Urbanization, Islamization and Crops on Pemba Island, Tanzania, AD 700–1500." *World Archaeology* 42 (1): 137–54.

Watson, Derek J. 2010. "Within Savanna and Forest: A Review of the Late Stone Age Kintampo Tradition, Ghana." *Azania* 45 (2): 141–74.

Watts, Michael J. 2013 [1983]. *Silent Violence: Food, Famine, and Peasantry in Northern Nigeria.* Athens: University of Georgia Press.

Wenger, Etienne. 1998. *Communities of Practice: Learning, Meaning, and Identity.* Cambridge: Cambridge University Press.

Whatley, Warren, and Rob Gillezeau. 2011. "The Impact of the Transatlantic Slave Trade on Ethnic Stratification in Africa." *American Economic Review* 101 (3): 571–76.

Widgren, Mats. 2017. "Agricultural Intensification in Sub-Saharan Africa, 1500–1800." In *Economic Development and Environmental History in the Anthropocene: Perspectives on Asia and Africa*, edited by Gareth Austen, 51–67. London: Bloomsbury Academic.

Widgren, Mats, Tim Maggs, Anna Plikk, Jan Risberg, Maria H. Schoeman, and Lars Ove Westerberg. 2016. "Precolonial Agricultural Terracing in Bokoni, South Africa: Typology and an Exploratory Excavation." *Journal of African Archaeology* 14 (1): 33–53.

Wilk, Richard. 2006a. *Home Cooking in the Global Village: Caribbean Food from Buccaneers to Ecotourists.* Oxford: Berg.

———. 2006b. "From Wild Weeds to Artisanal Cheese." In *Fast Food/Slow Food: The Cultural Economy of the Global Food System*, edited by Richard Wilk, 13–30. Lanham, MD: AltaMira Press.

Wilks, Ivor. 1975. *Asante in the Nineteenth Century: The Structure and Evolution of a Political Order*. Cambridge: Cambridge University Press.

———. 1977. "Land, Labour, Capital and the Forest Kingdom of the Asante: A Model of Early Change." In *The Evolution of Social Systems: Proceedings of a Meeting of the Research Seminar in Archaeology and Related Subjects Held at the Institute of Archaeology*, 487–534. London: Duckworth.

———. 1993. *Forests of Gold : Essays on the Akan and the Kingdom of Asante*. Athens: Ohio University Press.

———. 2004. "The Forest and the Twis." *Transactions of the Historical Society of Ghana*, n.s. 8: 1–81.

Wolf, Eric Robert. 1982. *Europe and the People without History*. Berkeley: University of California Press.

Worboys, Michael. 1988. "The Discovery of Colonial Malnutrition between the Wars." In *Imperial Medicine and Indigenous Societies*, edited by David Arnold, 108–225. Manchester: Manchester University Press.

World Food Summit. 1996. *Rome Declaration on World Food Security*. Available at http://www.fao.org/3/w3613e/w3613e00.htm.

Wutich, Amber, and Alexandra Brewis. 2014. "Food, Water, and Scarcity: Toward a Broader Anthropology of Resource Insecurity." *Current Anthropology* 55 (4): 444–68.

Wylie, Alison. 1985. "The Reaction against Analogy." *Advances in Archaeological Method and Theory* 8: 63–111.

———. 1989. "Archaeological Cables and Tacking: The Implications of Practice for Bernstein's Options beyond Objectivism and Relativism." *Philosophy of the Social Sciences* 19 (1): 1–18.

Yarak, Larry W. 1979. "Dating Asantehene Osei Kwadwo's Campaign against the Banna." *Asantesem* 10: 58.

York, R.N. 1965. "Excavations at Bui: A Preliminary Report." *West African Archaeological Newsletter* 3: 18–21.

Žižek, Slavoj. 2008. *Violence*. London: Profile.

INDEX

A-212 (site), 82, 178–79

Acemloglu, Daron et al., 28

Africa: assumption of development projects that lack of food is the problem to be solved, 7; "baseline," 12; denial of modernity for, 12; deficiency of large-seeded grasses in, 40; difficulty of writing food history for, 14–15; environmental determinists formulations of, 8; lacked *haute* cuisines, 133; as static and unchanging, 12-13, 130, 158-159

Africans: assumption of passivity of, 7; claimed to have poor diet by their choice, 117; currencies of, 127; food traditions not taken seriously, 6–7; recasting as inferior necessary in the time of slavery, 5; stereotypes about border on willful amnesia, 131; taste preferences differ from Westerners, 133

agriculture: cash cropping as move from moral to market economy, 126; colonial period documentation of, 30–31; cycle of, 111, 120–22, 136*fig.*, 137–38; "to eat and to sell" strategy in, 135–36, 137; emergence of, 25–26, 187–88n2; in forest savanna mosaic, 64–65; gender roles in, 109–10, 139; hiring of farm labor in, 139; importance of storage in, 25–26; intellectual history of, 23–24; intensive, 9; Iron Age, 26; labor organization and food redistribution are among Africa's chief developments in, 98; lack of archaeological interest in colonial period, 27; presence in forested areas may have been unrecognizable to Europeans, 64–65; recent shortening of fallow in, 137; rotation of crops, 137–38; use of mounds in, 137–38

Akan, 63

Al-Omari, 57

Alpern, Stanley B., 90

American crops: alleviate food security concerns today, 39; Big Bang Thesis about, 62–65; idea that they fueled rise of complex polities, 61; idea that they made up for population losses to slavery, 7, 61, 90, 91, 96; importance in Africa of, 1, 40; had minor role in eighteenth and nineteenth century Asante, 73–74, 93, 94. *See also* cassava; maize

archaeobotany: data, 35–36, 177–81

Asante, 61; annexation into Gold Coast of, 100; emergence of, 62–63; expansion and spatial organization of, 70–72; feasts, 76–77; food security, 78–79; geopolitical divisions of, 72*map*; historical scholarship on, 68; interpretation of lineage names of, 65; minor role of American crops in in eighteenth and nineteen century, 73–74, 93, 94; origin myths, 63; paramount chief, 62, 76–77, 78; polities and roads, 71*map*; power waned with end of slave trade, 86, 89; role of maize for, 16, 41, 73–75, 76; slaves, 69, 77; trade, 68, 69–70; tribute to, 72–73, 83, 93

Founded in 1893,
UNIVERSITY OF CALIFORNIA PRESS
publishes bold, progressive books and journals
on topics in the arts, humanities, social sciences,
and natural sciences—with a focus on social
justice issues—that inspire thought and action
among readers worldwide.

The UC PRESS FOUNDATION
raises funds to uphold the press's vital role
as an independent, nonprofit publisher, and
receives philanthropic support from a wide
range of individuals and institutions—and from
committed readers like you. To learn more, visit
ucpress.edu/supportus.